HOW TO GET RID OF THE

POISONS IN YOUR BODY

HOW TO GET RID OF THE
POISONS IN YOUR BODY

Gary Null and Steven Null

AN ARC BOOK

ARCO PUBLISHING, INC.
219 PARK AVENUE SOUTH, NEW YORK, N.Y. 10003

Fourth Printing, 1981

An ARC Book

Published by Arco Publishing, Inc.
219 Park Avenue South, New York, N.Y. 10003

Library of Congress Cataloging in Publication Data

Null, Gary.
    How to get rid of the poisons in your body.

    Includes bibliographical references.
    1. Nutrition 2. Food additives. 3. Cosmetics.
    4. Drugs. I. Null, Steve, joint author. II. Title

TX355.N86   615.9   76-28699
ISBN 0-668-04114-5 (Paperback Edition)

Printed in the United States of America

Acknowledgments

The preparation of this book required approaching the problem of what to say and how to say it. Several points we make here have never been published in book form before. Also, we felt it necessary to reiterate the problems, even though they have been stated a thousand times before in many excellent books. We forget that there is always a new audience, seeking clarity and guidance on this old problem of food and drug safety. Many of the important books, which first heralded the news of the dangers of sodium nitrite, Red Dye #2, DDT, etc., have long since gone out of print or are generally unavailable. It is not essential to open up an entirely new battle with every new book, especially if old battles have not been won. We therefore decided to try to reeducate the newly awakened consumer as to the nature of the controversy and all of its ramifications and effects upon the consumer. In addition, we offer a plan for consumer action and show that everyone can do something about his or her own state of health.

To prepare the data for this book required several years of investigating the facts and claims and charges made by both industry and consumer groups. We did not go into this with any predetermined conclusions. Therefore, we honestly feel that the conclusions drawn are objective and have not been motivated by political or economic interests. Unlike many "paid" consultants for the food and drug industries (including Health Food Consultants), we have never accepted any money or special favors or other forms of payment from any industry group.

During the past three years we have conducted several original research experiments in our laboratories and some of the results of these experiments are in this book. In addition, we interviewed hundreds of spokesmen and experts from all sides of the public health science field.

We have prepared a special list of reference books which can help any consumer and should be in every concerned citizen's home library.

GARY AND STEVE NULL

Consumer Beware, by Beatrice Trum Hunter, published by Simon and Schuster, New York, 1971.

Unfit for Human Consumption, by Ruth Mulvey Harmer, published by Prentice-Hall, Englewood Cliffs, N.J., 1971.

Food Pollution, by Gene Marine and Judith Van Allen, published by Holt, Rinehart, Winston, New York, 1972.

Eater's Digest, by Dr. Michael F. Jacobson, published by Anchor, New York, 1972.

The Great Drug Deception, by Ralph Adam Fine, published by Stein and Day, New York, 1972.

Bio-Organics, by James Rorty and N. Philip Norman, M.D., published by Lancer, New York, 1956.

Stop Poisoning Yourself, by Dr. Geoffrey T. Whitehouse, published by Award, New York, 1968.

Regulating New Drugs, edited by Richard L. Landau, published by The University of Chicago Center for Policy Study, Chicago, 1973.

Selling Death, by Thomas Whiteside, published by Liveright, New York, 1970.

Feast or Famine, by Ed Edwin, published by Charterhouse, New York, 1974.

Sowing the Wind, by Harrison Wellford, published by Bantam, New York, 1973.

The Chemical Feast, by James Turner, published by Grossman, New York, 1970.

The Dictocrats', by Omar Garrison, published by Arco, New York, 1970.

Silent Spring, by Rachel Carson, published by Houghton Mifflin, Boston, 1962.

A Consumer's Dictionary of Food Additives, by Ruth Winter, published by Crown, 1972.

The Brain Bank of America, by Philip Boffey, published by McGraw-Hill, New York, 1975.

Eating May Be Hazardous to Your Health, by Jacqueline Verrett, and Jean Carper, published by Simon and Schuster, New York, 1974.

Hungry for Profits, by Robert J. Ledogar, published by IDOC, New York, 1975.

Hot War on the Consumer, by David Sanford, Ralph Nader, James Rideway, Robert Coles, published by Pitman, New York, 1969.

The American Food Scandal, by William Robbins, published by Morrow, New York, 1974.

The Medicated Society, by Samuel Proger, M.D., published by Macmillan, New York, 1968.

Sweet and Dangerous, by John Yudkin, M.D., published by Bantam, 1972, New York.

Sugar Blues, by William Dufty, published by Chilton, Radnor, Pa., 1975.

The Saccharine Disease, by T.L. Cleave, published by Keats, New Canaan, Conn., 1974.

More Than Skin Deep, by Thomas Sternberg, M.D., published by Doubleday, New York, 1970.

Cosmetics, by Toni Stabile, published by Ballantine, New York, 1973.

Contents

HOW TO GET RID OF THE

POISONS IN YOUR BODY

Poisoning Your
Personal Environment

Most Americans buy their food in supermarkets where long rows of shelves are filled to overflowing with an endless variety of edible goods. With nature's bounty so plentifully and easily available, one might assume that Americans would be the healthiest people in the world. Unfortunately, this is not true. At least half the population suffers from some form of chronic illness such as cancer, heart disease, diabetes, high blood pressure, digestive problems, headaches, allergies, and low resistance to infections.[1] According to the World Health Organization, the United States ranks twelfth in general health—a shockingly low rating which clearly indicates that the people of poorer nations feel and look better than we do.

How can this happen in the most fertile country on earth, the "world's breadbasket," which produces such enormous quantities of food it feeds much of the rest of the world with its surplus?

The answer, while obvious, is a complex one.

Almost all of the food you buy—whether boxed, canned, frozen, plastic-wrapped, or so-called "fresh"—is saturated with chemicals that color or preserve it, or are in some other way intended to enhance its looks, crispness, or flavor. Making food look like something it is not is an important part of the $144 billion[2] food industry. The deception permits manufacturers to endow poor quality

9

raw materials with the misleading, up-graded appearance of better grade products. And the deception is at the risk of the consumer. Even small amounts of chemical additives ingested over a period of time can gradually harm the function of healthy tissue and organs.

The most insidious fact is that food contains many hidden additives that are not listed on the label. Pesticide sprays are absorbed by fruits and vegetables while they grow, and then are passed on to the consumer who tends to believe that anything fresh must be uncontaminated. In addition, hormones and antibiotics are administered to livestock and poultry to increase their yield and make them grow fatter, which means that meat eaters are being dosed with these physiologically active drugs. Cosmetics and drug store remedies are also laced with chemical compounds to extend their shelf life or to impart a pleasant scent; some are given attractive color with cancer-causing coal-tar dyes.

Since the consumer does not immediately have a serious reaction upon ingestion of these toxins, but instead exhibits a variety of symptoms that are almost always mistaken for something other than body pollution, he is lulled into a false sense of security. Large doses of these chemicals *can* kill you on the spot, but the food industry watchdog, the Food and Drug Administration (F.D.A.), limits their use to what they consider "safe" human tolerance levels. By doing so, our government is actually sanctioning the slow mass poisoning of its citizens.

Our Sad State of Health

Our health agencies point with pride to the fact that the average American has a life expectancy of approximately 72 years,[3] which is double that of his or her ancestors of a century ago. But this statistic is misleading and becomes less impressive under close examination. One hundred years ago many more infants died than do today, which reduced the *statistical* average life expectancy of the entire population of the United States, not the life expectancy of individual adults. The truth is, a nineteenth-century American who survived to the age of 40 had a life expectancy of only two years less than today's. And our ancestors who reached 60 could expect to continue living longer than today's 60-year-olds. In other words, we have increased our average life span through the excellent work done by the medical profession in delivering more healthy babies and conquering common childhood diseases, i.e., scarlet fever, diphtheria, and dysentery.

That we have been able to increase a mature individual's actual life span by only two years in a century—a century of amazing advances in surgical techniques, public health measures, and drugs that pull seriously ill persons from death's door—points to a great failure in keeping our general health at a high level commensurate with the technological means available to us. During the years that average life expectancy was increasing, the adult death rate from degenerative diseases was rising dramatically—a rise that parallels the increasing consumption of processed, chemically treated foods. More of us than ever before are dying of cancer, heart disease, and kidney disease. A look around any drug store tells us that many Americans are suffering from a variety of vague symptoms for which they take non-prescription medications which only make matters worse. Patent medicines do not offer cures; they only mask pain, preventing the user from getting to the source of the malady, while adding their own pollution to the body.

There is further clear evidence which tells us there is something radically wrong with the American health:

—The American Heart Association reports that in the United States heart disease accounts for one-third of all deaths, making us the world leader.

—Twenty-three countries have a lower infant mortality rate than we do.

—8,000 to 10,000 apparently normal babies die each year from sudden "crib death," which medical science is able neither to explain nor to control, much less to prevent.

—Rejection of armed forces volunteers for physical and mental deficiencies is well above the 40 percent rate.

—In an examination of 10,000 highly placed business executives, doctors found only one in ten to be in general good health.

—In a ten-year period, the number of patients admitted to hospitals increased at a rate five times faster than the growth of the population.

—One family in seven has a member who undergoes surgery in any given year.

It is the purpose of this book to alert the consumer to what has been done to his or her food, cosmetics, and drugs, and to offer sensible alternatives to body pollution.

The Land of Synthetics

Although all the foods we need for a healthy, fulfilled life already exist in nature, the food industry wants us to believe that "improved" and "enriched" products are essential to our well-being. These foods are actually inferior, mass-produced preparations built around a framework of synthetics designed to titillate the tastebuds while shortchanging the metabolism. Both food processors and government agencies maintain policies of secrecy, deliberately concealing from us the inferior quality of the food we eat. We are not told that the hundreds of additives present in our food are hazardous rather than beneficial to our health.

On the assembly line, fruits and vegetables are rinsed of pesticide residues which, however, remain *inside* the plants with detergents, solvents, and oils. They come into contact with adhesives, resins, plasticizers, defoamers, textile fibers, and rubber catalysts used in packaging. Most of these remain in the food even after it is scrubbed and cooked. These chemicals are not listed on labels as additives, of course, because they were not "added during the processing and packaging."

The food industry sells products by making them taste good and look attractive with chemical camouflages that hide lack of purity and wholesomeness. For example, the industry has spent millions of dollars advertising TV dinners as a nutritious time-saver for the busy homemaker—with great success. Yet, the contents of the package are vastly different from the succulent-looking array depicted on the container. The chicken is hidden deep inside a thick coating of batter, and the illusion that it has been cooked to a tempting doneness has been achieved with artificial coloring, synthetic flavor, and chemical preservatives. Most of what looks like meat is a cheap batter filler composed of flour, sugar, and salt.

High-carbohydrate foods like TV dinners, instant potatoes, soup mixes, and cake mixes are very popular with manufacturers because they keep well and are inexpensive to produce, making them ideal for mass distribution. These convenience foods often cost less to produce than their advertising and packaging budgets. The fact that they lack vitamins, minerals, and amino acids characteristic of first-rate foods is concealed behind a facade of glossy packaging and advertising falsehoods that appeal to our emotions rather than to our good sense.

Although few people live exclusively on convenience foods, the problem does not end there. To millions of Americans, processed cheese slices and spreads are synonymous with cheese itself; they do not realize that "American cheese" and "Cheez-whiz" are the TV dinners of the cheese world. According to the U.S. Department of Agriculture's industry oriented booklet, *Cheese Varieties*, "Process cheese is made by grinding fine, and mixing together by heating and stirring, one or more cheeses of the same or two or more varieties together with an emulsifying agent, into a homogeneous plastic mass. Lactic, citric, acetic or phosphoric acid or vinegar, a small amount of cream, water, salt, color, and spices or flavoring materials may be added."

This booklet neglects to tell us the manufacturer's reason for inventing this process: Could it be to use up stale and defective cheese? Decayed rinds, pinholes, and other unattractive remains could all go into the process cheese. The emulsifier may contain as many as twelve different chemicals, and a chemical stabilizer may be added as well. Of course, the cost of all these nutritionally empty chemicals is passed on to the consumer.

The hot dog, which was once a decent food, has undergone a similar fate. Forty years ago, this popular food consisted of approximately 20 percent protein and 19 percent fat—about the same proportions as those found in meat. During the 1950s, however, manufacturers discovered ways to add more fat and water to the product, without causing the ingredients to separate. The average frankfurter on the market today contains less than 12 percent protein, some of which consists of chicken by-products. It also contains about 30 percent fat, and 50 percent water. The remaining 8 percent consists of filler, salt, sugar, sodium nitrite, sodium nitrate, sodium erythorbate, and a red coal-tar dye.[4] Thus, today's consumer pays a high price for a frankfurter which is chemically treated, and not nearly as nutritious as it was during the Depression.

Wasted Nutrition

The problem of body pollution is further compounded by the fact that modern food processing methods remove many vitamins and unknown nutrients that might help protect us from additives. When a food is divided into its component parts either by refinement or extraction—such as the separation of all cream and butter from milk— nutritional losses are inevitable. The trace minerals of magnesium, copper, cobalt, chromium, and molybdenum remain in the cream and butter, while the manganese and zinc stay in the skim milk. Homogenization was invented to solve the marketing problem of cream gathering in large globules at the top of milk. In this process, the cream is combined with butter and then put back into the skim milk. This way the dairy farmer is never stuck with surplus butter; it can always be mixed back into fresh whole milk adulterated with water and skim milk powder. It is now virtually impossible to find anything but homogenized milk in a commercial outlet—milk that is in effect reconstituted, its molecular structure changed so that it will not separate. In the homogenizing process, several vitamins and minerals are lost, but only A and D are replaced.

Often, the milk we buy seems to have an off flavor or it spoils sooner than the date stamped on the container. This is because although the milk has been cooked at a high temperature—pasteurized—to destroy disease-producing organisms, some bacteria are not killed and multiply at a fantastic rate during the time it takes for the milk to be brought to market. Some of this contamination comes from fecal matter on the cow's flanks, teats, and udder, and is called coliform bacteria. Rather than insist that milk be brought to market sooner and cleaner, the F.D.A. allows a coliform bacteria count to be taken by the processor and not at the point of sale to the consumer. Therefore, the allowable count of coliform bacteria is meaningless since the bacteria is sure to multiply by the time we drink the milk.[5] Most of us don't realize that in effect we have been poisoned; stomach aches and diarrhea are usually attributed to other causes.

The dairy industry justifies its mass-production methods by pointing out that there is less waste than before, and that milk and dairy products are cheaper to produce. Cheaper for whom? As usual, the industry is talking about its own advantage, not ours; when homogenization was introduced, milk prices went up, not down.

The molecular modification of vegetable oils for use in margarine and shortening is another example of how a health-promoting food can be turned into an inferior product. To make them solid and butter-like at room temperature, vegetable oils rich in polyunsaturated lipids (which help keep our arteries supple and free-flowing) are subjected to high temperatures and pressure, a process which destroys most of the oils' vitamins and natural therapeutic qualities. Hydrogen is pumped through the oil in the presence of nickel or another metal catalyst, which when absorbed, makes the oil hard. The hydrogenated oil is a dark, foul-smelling grease which is then bleached, filtered, and deodorized into a white, lard-like fat that must be artificially colored and flavored before being marketed as margarine. Fat-soluble vitamins such as A, D, K, and E are destroyed in the process, and the essential fatty acids (E.F.A.) factor is converted into an abnormal form that many researchers believe to be antagonistic to the body. Modern science has made it possible to change beneficial oils into cholesterol-producing artery hardening saturated fats.

All processed foods are prepared under similarly severe conditions of temperature, agitation, and pressure which rob them of their original nutrients. Prodded by public pressure and the F.D.A., some food manufacturers now "fortify" their inferior products by restoring a few of the vitamins and minerals that have been destroyed. Potatoes processed as flakes to be reconstituted perhaps years later often lose more than 90 percent of their vitamin C content, making it necessary to add synthetic ascorbic acid to the product.[6] Similarly, white flour and white rice are stripped of most of their nutrients—including bowel-cleansing roughage factors during the harsh milling process—and little more than starchy empty calories remain. A few nutrients are added later in a futile attempt to cover this great loss. The word "enriched" printed on a bag of white flour or rice is misleading for the consumer usually reads the word to mean that a natural product has been improved, which is obviously what the industry intends.

Most of us do not think of vitamins and minerals as food additives, but in a sense they are, especially when the vitamins are artifically produced. All the elements of proper nutrition are in balance in natural, organic foods, a balance that is completely cancelled in the manufacturing process. Scientists assume that there are a number of as yet unidentified vitamins and trace elements that are essential to the body, and that these act in conjunction with those we know about. To replace only a few of these key nutrients while neglecting the level of others intimately associated with them, can result in serious deficiencies.

Cadmium is always present in a food substance containing zinc. If zinc is absent, cadmium rushes into take its place—and too much cadmium is hazardous to human health. In whole wheat and whole wheat flour, zinc and cadmium are kept in a natural, opposing balance, which is expunged when flour is refined and bleached. Iron is usually restored to white flour, but the copper that is needed for its proper utilization is not, which means that the body will not absorb it.

In 1970, Dr. Roger J. Williams, an eminent biochemist of the University of Texas, told a meeting of the National Academy of Sciences of an experiment he had conducted with 64 mice that were fed only bread made from refined flour. Within 90 days, 40 of the rats had died of malnutrition.

"I see no reason why," Dr. Williams told the panel, "in an age of scientific advance, the product of a modern bakery should not far surpass whole wheat bread in nutrition." Dr. Williams later angered the baking industry when he noted that "today's bread has about the same nutritional value as sawdust." Up in arms, the industry countered that bread is not usually consumed alone, which prompted a further rebuttal from Dr. Williams: "Sawdust, when accompanied by good food (milk, meat, and cheese) can yield acceptable results, yet sawdust is known to be devoid of nutritional value."[7]

Some breakfast cereals are major body polluters. Once-healthful grains such as corn, rice, wheat, and oats are steamed, pressurized, toasted, flaked, puffed, coated with sugar (itself a refined, pure chemical), and preserved with BHA (butylated hydroxyanisole) and BHT (butylated hydroxytoluene), two petroleum products used as antioxidants. Then as an extra touch, they are laced with a few vitamins, and advertised as "wholesome and nutritious." In reality, they are little more than vitamin coated candy that could cause more harm than good to the person who eats them.

In the summer of 1970, consumer advocate Robert Choate proved that the "Breakfast of Champions" was only an advertising slogan, when he conducted laboratory tests that showed dry cereals to be lacking in body building protein and other nutrients essential to health. The cereal industry responded to Mr. Choate's findings by tossing in a few more "enrichments" and stepping up their advertising campaign on children's TV shows. "The kids are going to be with us longer," explains the head of Kellogg; a more plausible explanation might be that children do not read newspapers, and they are more likely to respond to the "other-kids-are-eating-it" appeal of commercials. As for the "enrichments," consumerist Michael F. Jacobson reports: "For their generosity in adding a half cent worth of vitamins to twelve ounces of cereal, they add 18¢ (45 percent) to the retail price."[8]

Fast Foods—Faulty Nutrition

Independent consumer studies show that one out of every three meals is eaten outside the home, and if the trend continues Americans will be eating half their meals in restaurants by 1980. Judging by the fast-food industry's phenomenal growth in the last 15 years to a $10 billion a year industry,[9] most of those meals will be eaten in drive-in, self-service restaurants that serve only a limited menu of additive-ridden nutritionally questionable foods.

Besides producing "visual pollution," fast-food chains could contribute to the decline in national health. The Department of Agriculture's dietary surveys showed recently that the number of Americans who received an adequate diet actually declined between 1955 and 1965—

the same period in which fast-food franchisers and processed food manufacturers measured their greatest growth. Dr. H. Curtis Wood, Jr., the noted nutritionist, writes in his book, *Overfed but Undernourished:* "Young people have an amazing ability to adjust and adapt to unbelievably deficient diets and yet feel fairly well. Because of this, they do not understand that while they may be able to get away with living on hamburgers, potato chips, and soda pop without apparent disastrous results for some years, it will eventually catch up with them in one way or another."

As the reader will see in a later chapter, sugar is a highly addictive pure chemical that creates a constant demand for more, a fact that is not lost on owners of fast-food restaurants where everything but the French fries are sweetened. Sugar is added to the hamburger bun, to the meat, to the pickles, the catsup, and the "special sauce." The milkshake is not worthy of its name, for it is made of water, sugar, thickening and stablizing agents, emulsifiers, preservatives, artificial coloring and flavor, and a tiny amount of powdered milk.

The total caloric count of a fast-food meal consisting of a full-sized hamburger, french fries, and milkshake is approximately 1,300, or one-half the daily ration of calories needed by an adult male. In other words, if a person ate such a meal regularly, while consuming two other meals each day, he or she would gain weight. Women and children would put on even more weight than men, since their caloric requirements are lower. All fast foods are presalted, and *Consumer Reports* magazine found that the average sodium content of the eight meals that were tested was well over a teaspoonful—and that does not include the table salt that most people sprinkle on their food!

Too much salt in the diet is a contributing factor to high blood pressure, a common ailment that can lead to dire consequences when compounded by an excessive intake of cholesterol-producing saturated fats and refined sugar, and smoking and drinking. Most of us need little more salt than that which is contained in food naturally.

Among the nutrients found lacking in fast foods are several vitamins from the B group—folacin, pantothenic acid, and biotin—which must be present for the other B vitamins to be completely absorbed. Several years ago, when pantothenic acid was first discovered, a group of dogs were fed a diet lacking this essential nutrient for three months. They appeared to be healthy enough up until a day or two before they suddenly died. What had happened was that the heart muscle had leeched pantotheic acid from every cell in the body, and then starved to death when there was no more left. Vitamin A, iron, and copper are also in short supply, which means that the excessive burger fancier may be plagued with night blindness, respiratory infections, and/or fatigue and anemia.

The Baby Food Flim Flam

If those who frequent fast-food chains were to eat at least two well-balanced meals daily—including whole grains, beans, dark green, leafy vegetables, yellow vegetables, and fresh fruits—the deficiencies caused by the fast foods could be overcome. However, most of those who prefer salty, high-carbohydrate diets are addicted to the taste of assembly line foods, and they simply eat more of them in other forms. The habit usually begins in infancy when parents launch their children into a lifetime of bad eating habits by feeding them commercial baby foods.

To begin with, baby food manufacturers add unnecessary amounts of water to stretch strained vegetables, meats, fruits, and custard desserts. An infant wouldn't notice that it is being cheated, but Mom and Dad would, so manufacturers add inexpensive modified starch to restore a thick, natural-looking texture to the food. It makes no sense whatsoever to add sugar and extra salt to strained fruits and vegetables, because baby can't taste them anyway. Nonetheless, manufacturers do so because they know that parents taste the food first and mistakenly believe that what tastes good to them will taste good to their baby.

Until recently, monosodium glutamate (M.S.G.) and sodium nitrite were also added to baby foods to please Mom and Dad with intensified flavors and colors. When the public found out that these brain-damaging and cancer-causing chemicals were routinely threatening the fledgling metabolisms of their infants, the outcry was so great that the F.D.A. was finally forced to ban their use in baby foods (but not in any others). While any government action against the use of harmful additives is to be applauded, the unfortunate result was that parents were then lulled into a sense of false security. The fact is, baby foods are still being adulterated in other ways.

By displacing from 10 to 30 percent of natural ingredients with water, starch, sugar, and salt, and by cooking the food at high temperatures that destroy many vitamins, manufacturers are producing a junior-sized fast-food meal. Babies fed only commercial baby foods simply may not receive the food needed for maximum development of bone, organs, skin, muscles, nerves, and other parts of the body. Researchers have found that baby foods containing starch and sugar at the expense of protein and unsaturated fats predispose the infant to becoming an overweight, sugar-addicted adult. You will be doing your child a favor by making your own baby food, and steering him or her into a proper dietary course early in life.

Foodless Foods

Most parents, however, find it easier to continue feeding their children more convenience foods as they grow up. A candy bar is a reward for good behavior; potato chips are packed in lunchboxes instead of apples; after-school snacks are cream-filled cupcakes rather than dried fruit or nuts. Sweetened, artificially colored, emulsified, and thickened "foodless foods" deficient in everything but carbohydrates become the norm, and the child consumes fewer of the raw, whole foods necessary for growth and health.

Recently, a group of graduate students at the University of Florida conducted an experiment which indicated that American youth actually prefer artificial foods. Given a choice of synthetic orange juice made from a powder mix, reconstituted frozen orange concentrate, and freshly squeezed juice, most of the teenagers surveyed preferred the sugar-sweetened imitation. Second best was the frozen product, which was familiar to most of them; but the fresh orange juice was judged to have a "weird" flavor. For a large number of them, it was the first time they had tasted the rich, pulpy original!

Margaret Mead, the renowned anthropologist, correctly describes processed edibles found in our supermarkets as "foods guaranteed not to nourish you." Young people, of course, often do not possess the wisdom to make wise choices, and they are easy prey for the nutritional misinformation presented on television advertisements for snack foods. Dr. Jean Mayer, the Harvard nutritionist who has been a White House advisor, accuses the food industry of selling-out the nation's health by promoting products beneficial only to themselves.

"If you classify foods from the most useful to the least useful," Dr. Mayer says, "you'll place meat, vegetables, fruits, milk, and eggs in the first category, bread, pastas, and cereals in the second category, snacks in the third, and candy, soft drinks, and beer in the fourth. But if you think about it, you'll realize that the bulk of the food advertising is for things in categories 3 and 4—the least useful foods."[10]

The inevitable results to a generation weaned on inferior foods—a generation that spends a great deal of time watching other people exercise their bodies on televised sporting events—was verified early in 1976. According to a study sponsored by the U.S. Office of Education, the physical fitness of American boys and girls underwent a decline between 1965 and 1975, reversing a 100-year trend. Simon A. McNeely, program director of the study, blames the decline on "the continuous inroads of soft living in the United States."[11] Explaining the scope of the problem, Canadian child nutritionist Dr. Abram Hoffer states that nutritional imbalances are often manifested as "overactivity, perceptual changes, difficulty in reading and learning, and changes in personality." Often such a child is mistakenly believed to be retarded or mentally ill. Misfed teenagers, says Dr. Hoffer, are poor students, are moody, and are inclined to take drugs like marijuana to make themselves feel better.[12]

The Fat American

Overeating has become one of the nation's most vexing health problems. Millions of Americans are bloated and made sick by the abundant inferior foods they consume. Deprived of the essential nutrients which protect against obesity, many of us plod along on imbalanced, improper diets, never achieving our full potential of mental and physical vigor. Overweight people have a mortality rate more than six times higher than those whose weight is normal; they are more likely to die from cancer and heart attacks; and their rate of diabetes and blood vessel disease is doubled.

Obesity is not a disease in itself, but a symptom that something has gone wrong with the self-regulating mechanism that balances intake and utilization of food materials. Obesity is simply a condition in which more calories are ingested into the body than are used up. The unused calories are stored in the body as fat. Ironically, obesity is considered a form of malnutrition in that the body receives too many of the wrong foods and not enough of the right ones, which contributes to a variety of ills.

Many nutritionists and medical doctors believe that the root cause of obesity is a high-carbohydrate childhood diet. The plump child, whose full, rosy cheeks are supposed to be indications of good health, has a greater than average chance of growing up to be a fat adult. Such children may develop an abnormal amount of extra-large fat cells that persist throughout life, making obesity a permanent condition. No matter how hard they try, these unfortunate individuals find it all but impossible to lose weight because their fat cells require large amounts of nourishment; chronic hunger almost always defeats their willpower. Parents could spare children this agony of obesity in adulthood by making wiser choices in the food they give them.

Anyone attempting a weight-reducing diet should first consult a physician to verify that the condition is not caused by a glandular disorder. The majority of people, however, will respond to a low calorie diet of protein and fresh vegetables, no starches and sugars, and not more than two slices of preservative-free, whole-grain bread each day. Meat should be limited in favor of fish, cheese, occasional eggs, and vegetable proteins, for the residue of female hormones given to livestock to make them fatter may have the same effect on heavy meat-eaters. Certain vitamins may be taken to supplement the diet and help reestablish the body's natural equilibrium. Eventually, the regimen of wholesome, health-promoting foods will balance the appetite-regulating mechanism, and you will gradually be able to add grains, cereals and fruits to your daily meals. You will find that it is difficult to eat more of the vitamin and mineral-rich carbohydrates than your body requires.

Equally important, getting rid of excess pounds also means that you are cleansing your body of chemical additives absorbed during many years of faulty eating. Since most of these manufactured body pollutants are stored in body fat—where they can interreact with each other and healthy cells to produce a number of familiar illnesses such as high blood pressure, cancer of the liver, and kidney ailments—it is imperative to maintain a normal weight to minimize the potential damage such pollutants may cause.

Damage to the Unborn

Among the most disastrous (and often tragic) changes that are caused in human tissue by chemical irritants are mutations and birth defects. Mutations occur when the genetic functioning of reproductive cells is altered and an unwelcome characteristic is passed along from generation to generation. This may be a lowered resistance to disease, a reduced lifespan, mental retardation, or congenital malformations. Some food additives are proven mutagens. They attack DNA (deoxyribonucleic acid, the fundamental component of all living matter) molecules contained in a woman's egg and a man's sperm, and scramble the instructions for the proper development of offspring. Genetic mistakes, ranging from misshapen or extra fingers or toes to such serious diseases as hemophilia, can result.

Dr. James F. Crow, chairman of the Genetics Department of the University of Wisconsin, warns that this factor must be taken into account in government tests to ensure the future of the human race. "Even though the compounds may not be demonstrably mutagenic at the concentrations used," Dr. Crow writes, "the total number of deleterious mutations induced in the whole population over a prolonged period of time could nevertheless be substantial. Such an increase in mutation rate probably could not be detected in a short period of time by any direct observation on human beings. Protection from such effects must depend on prior identification of mutagenicity."[13]

While some birth defects are caused by mutations, others are caused by chemical agents that induce changes in the developing fetus. These chemicals, which are known as teratogens, do not cause the defect to be passed on to future generations. The National Foundation—March of Dimes estimates that each year about 250,000 American babies are born with such imperfections as club foot, open spine, and cleft palate ("harelip"), and that 80 percent of these are caused by food additives, environmental pollutants, drugs, and alcohol.

A well-known and tragically dramatic example of a teratogen is thalidomide, a drug intended as a sedative and sleeping aid that deformed the fetuses of mothers who used it during the first trimester of pregnancy. In the early

1960s the epidemic of European babies born with a seal limb (flipper-like appendage technically known as phocomelia) proved once and for all that a fetus can be harmed by substances foreign to the body, even though the mother is unharmed—a fact which was formerly considered a test of safety.

Meanwhile, those afflicted with congenital defects must go through life shouldering the burden of the carelessness of previous generations. In addition to the grief caused, the cost in dollars is tremendous. The victim's family and government social agencies must pay for medical treatment, hospital rooms, and homes for the retarded and seriously physically afflicted, which can cost as much as $100,000 per person annually. Not included in this figure are loss of earnings and failure to contribute to the economy. It would not be difficult to eliminate most of these needless tragedies by keeping our food supply pure and natural and by staying away from harmful drugs, cosmetics, cigarettes, and alcohol. Sadly, government agencies like the F.D.A. have done little to screen out dangerous chemicals that adversely affect our genes.

Don't Let It Happen to You!

The debasement of the American diet has been made possible by the misuse of modern technology and chemistry—and by our willingness to let it happen to us. Processed foods are thought to be necessary in an industrial, mass-production society, but their use actually creates more problems than it solves. Urged on by persuasive, high-powered advertising, many of us are "nutritional illiterates" who select contaminated, nutritionally inferior food which greatly increases our risk of illness and that of our children. I hope the following pages will serve as your guide to optimum nutrition and freedom from body pollution.

Notes

1. Consumer Bulletin Annual, 1972.
2. U.S. Bureau of Economic Analysis, 1973.
3. U.S. Center for Health Statistics, *Vital Statistics of the U.S.*, Annual, 1975.
4. Consumer Reports, *Frankfurters*, February, 1972.
5. Consumer Reports, *Milk: Why Is the Quality So Low?*, January, 1974.
6. Herber, L., *Our Synthetic Environment*, Knopf, New York, 1972.
7. Williams, Roger J., *Nutrition Against Disease*, Pitman, New York, 1971.
8. Jacobson, Michael F., *Eater's Digest*, Doubleday-Anchor, Garden City, New York, 1972.
9. Consumer Reports, *How Nutritious Are Fast Food Meals?*, May, 1975.
10. Spencer, Steven M., "Are We Overfed but Undernourished?," *Contemporary, The Denver Post*, August 1, 1971.
11. U.S. Office of Education, March, 1976.
12. Ebon, Martin, *Which Vitamins Do You Need?*, Bantam Books, New York, 1974.
13. Crow, Dr. James F., "Chemical Risks to Future Generations," *Scientist and Citizen*, June/July, 1968.

Sweet Suicide—
Sugar and Sugar Substitutes

Sugar is exactly as "pure" as the label on the bag proclaims. Its sparkling white crystals are clean and sterile, having had every trace of anything nutritious extracted from them in the refining process. Sugar has no vitamins, fats, minerals, proteins, or any other nutrient essential to your health. The only thing sugar (sucrose) offers the body is quick energy from the calories it contains. It is an unadulterated chemical that has been linked to diabetes, dental caries, obesity (being a major part of refined carbohydrates), and stomach problems.

But sugar tastes good. In fact, it is one of the most pleasant ways imaginable to do yourself in gradually. Sugar works its harmful ways very slowly, and the victim rarely becomes aware of the damage wrought until it is too late to reverse or even arrest it. Last year Americans willingly poisoned themselves by swallowing an average of more than 130 pounds of sugar[1] in any number of nutritionally inferior foods, thus making sugar a possible contributing factor to escalating illness rates.

The U.S. Government figure translates into approximately 31 teaspoons of sugar per day for every man, woman and child. Since there are many people, like ourselves, who do not consume processed sugar at all, and many others who prefer equally dangerous artificial sweeteners, the statistic is misleading, since it was arrived at by simply dividing the amount of sugar sold last year by the total population of the United States. Obviously, a great number of Americans are ingesting more than their share of the national average. A glance at the average middle-class American's diet reveals a more accurate profile.

Breakfast usually begins with a glass of sweetened canned or frozen fruit juice. Cereal is either sugared at the table or purchased presweetened. Toasted bread, muffins, waffles, or pancakes also contain sugar, as does the jam, jelly, marmalade, and syrup spread on them. Add one or two teaspoonfuls of sugar stirred into coffee or tea: *10 teaspoons of sugar*.

A midmorning snack is probably a sweet Danish pastry or doughnut, washed down by a sweetened cup of coffee or tea: *6 teaspoons of sugar*.

Lunch may consist of canned vegetable soup, a baked ham sandwich with pickle, followed by apple pie and another beverage, all of which contain sugar: *8 teaspoons of sugar*.

A midafternoon candy bar or cola soft drink "for energy" adds more empty calories: *6 teaspoons of sugar*.

At dinner, more sugar than ever is consumed. Cocktails and wine contain this noxious chemical, as does mayonnaise on the salad, the canned meat gravy, the frozen vegetable, and the hot rolls or biscuits. Then a piece of chocolate cake for dessert: *14 teaspoons of sugar*.

Later in the evening, the television advertisers work their subtle ways to stimulate the viewer's "sweet tooth," and send him or her scurrying into the kitchen for yet another soft drink, candy, cookies, ice cream, or another helping of dessert: *6 teaspoons of sugar*.

It is clearly evident that in fact the total amount of sugar consumed by an average American is more likely to be *50 teaspoons,* not 31. And even this figure is conservative, for it does not include popular American confections such as milkshakes, sodas, sundaes, and the heavily sugared foods available at "fast-food" restaurants, which feed an ever-growing number of malnourished Americans. Neither does it include white flour, another source of empty calories, which is converted to sugar in the intestine.

The body struggles valiantly to combat the damage done by consuming so much sugar while the person sleeps, but to no avail. The vicious cycle will be repeated over countless tomorrows until the body finally breaks down.

Fifty teaspoons of sugar represent more than 800 calories, approximately one-third of the body's daily requirement of food energy. Pure carbohydrates like refined sugar and flour cause a serious disruption in the body's metabolism by providing cells with energy, without feeding them nutrients essential for their proper functions. For the sake of clarification, let's imagine that the human body is an automobile engine. If you put only gasoline (sugar) into the car and neglect to lubricate it with oil (vitamins, minerals, proteins, and fats), the engine will soon break down. That is what happens to your body when you displace nourishing foods like whole grains, cheese, raw vegetables, and fruits with a chemical having no essential nutrients whatsoever.

In order to utilize refined carbohydrates, the body must rob healthy cells of nutrients they need to survive. To digest sugar, the body leaches precious vitamins and minerals from itself, inducing a crisis state. Trying to restore an acid-alkaline balance to the blood, the metabolic system draws sodium, potassium, and magnesium from various parts of the body, and calcium from the bones. Often, so much calcium is depleted that the result is brittle, easily fractured bones, and soft teeth susceptible to decay. Glutamic acid and other B vitamins are actually destroyed by the presence of sugar in the stomach, a condition evidencing itself in heavy sugar users by fuzzy thinking and a tendency to become sleepy during the day.

Often, there are not enough vitamins, minerals, and proteins such as enzymes present in the body to complete the conversion of sugar into energy. Consequently, the carbohydrates are incompletely metabolized, leaving residues such as lactic acid. These poisons accumulate in the brain and throughout the nervous system, where they deprive cells of oxygen. Eventually, the cells die, the result being that the body degenerates and becomes more susceptible to disease.[2]

Sugar manufacturers urge us to "eat sugar for quick energy," successfully luring nutritionally ignorant people into "sweet suicide." True, sugar does raise the blood-sugar level temporarily, only to let it fall rapidly (if not abruptly) a short time later. Sugar is absorbed into the bloodstream through the intestines within minutes of being ingested, producing a rush of "quick energy." But half an hour later the sugar is used up and a person is left with the familiar symptoms of hypoglycemia (low blood sugar), headache, dizziness, fatigue, and irritability. The common reaction is to reach for something sweet and repeat the process.

Hypoglycemia has become a familiar complaint in our sugar-saturated age, but few know how—or have the willpower—to overcome it. When we eat, about 68 percent of *all* food, provided there is ample protein, is converted into glucose, a simple sugar that is almost identical to white sugar (sucrose) which supplies us with energy. The remaining 32 percent is used for repairing and building the body cells. Not only carbohydrates but also fats and proteins can be transformed by the body into the type of sugar it needs to generate energy. To cure low blood sugar you must replace white sugar and flour with a diet rich in slow-burning proteins and natural carbohydrates—fruits and vegetables—plus a moderate amount of foods containing natural fats, such as soybean oil and avocados.

Refined sugars and starches create havoc in the body's self-regulating mechanisms. Ordinarily, sugar is converted into glucose by the digestive juices and is carried through the bloodstream to the pancreas, which is stimulated into producing insulin. Excess sugar, in a more complex form known as glycogen, is stored in the liver for future use. When the blood sugar level begins to drop—a natural occurrence between meals—hormones signal the liver to convert some of its stored glycogen back into glucose. This reconstructed glucose is fed directly into the bloodstream, restoring a balance to the blood sugar level. In a properly functioning body, a healthy blood sugar level is maintained by the automatic interaction of insulin and hormones secreted by adrenal and pituitary glands.

Eating an excessive amount of sugar overstimulates the pancreas into producing an excess of insulin, and several things may happen. Most frequently, the person will develop an addiction to sweets to satisfy the insulin secretion. Or, the insulin will begin to convert too much glucose into glycogen and the blood sugar level will drop dangerously low, creating a condition which results in chronic hypoglycemia.

Surplus sugar is also stored by the body as fat. When the liver has all the glycogen it can absorb, the excess glycogen is spilled into the bloodstream in the form of fatty acids. Transported throughout the body, these liquid fats are deposited in least-exercised areas such as the buttocks, stomach, thighs, and breasts. When these areas can absorb no more, the fat is stored in major organs such as the kidneys and heart, and causes a decrease in their functions, eventual physical deterioration, and results in severe damage to tissues throughout the body.

Obesity is directly related to over-consumption of refined carbohydrates, and both cause and effect are widely held to be a prime factor in diabetes. The opposite of hypoglycemia, diabetes results from an inadequate supply of insulin and too much sugar in the blood. Exhausted from the constant demand of producing insulin to convert all that sugar into heat and energy, the pancreas will finally malfunction and the excess sugar then pollutes the bloodstream. Without sufficient insulin to convert the sugar into glucose, the body is deprived of an essential food, and the diabetic remains hungry, no matter how much he eats. Sugar accumulates in the bloodstream faster than the body can excrete it through the kidneys in the urine, literally poisoning the victim. He becomes tired, weak, and nauseated, can go into a diabetic coma, a condition which will result in death unless insulin is immediately administered.

Rather than instruct their diabetic patients to eliminate refined sugar, candy, soft drinks, and other sweets from their diet, the medical establishment latched onto an imperfect alternative. After years of research, Canadian Doctor Frederick Banting discovered a way to extract insulin from animals in slaughter houses. Dr. Banting received a Nobel Prize for his work in 1923, and drug companies celebrated one of the first "miracles" of modern medical science, eager to exploit a captive market of a million or so diabetics,[3] a market that was destined to grow. As William Dufty points out in his valuable book,

Sugar Blues, Dr. Banting later "tried to tell us that his discovery was merely a palliative, not a cure and that the way to prevent diabetes was to cut down on 'dangerous' sugar bingeing."

"In the U.S. the incidence of diabetes has increased proportionately with the per capita consumption of sugar," Dr. Banting later wrote. "In the heating and recrystallization of the natural sugar cane, something is altered which leaves the refined products a dangerous foodstuff."[4]

Discovered by Western traders in India during the fifth century, sugar cane did not come into general use until the seventeenth century, when colonists discovered that it could be grown cheaply and in profusion in the New World. Coffee, tea, and cocoa owed their growing popularity to the new abundance of sugar, and inventive chefs were inspired to create delicious new confections saturated with sugar. Wars often cut off Europe's supply of sugar from the West Indies, so enterprising chemists discovered a way to refine sugar from beet juice, which produced as pure a chemical as cane.

Since sugar in its natural state is a vital component of plants—sugar cane and sugar beets—the raw material does contain valuable nutrients. In the refining process, however, *all* that is beneficial is destroyed. After the cane is harvested, it is chopped into small pieces which are crushed by huge rollers. Water is added and the juice is squeezed out, filtered, and poured into heating vats. Powdered lime is then added to separate and coagulate most of the extraneous matter.

The heated brownish liquid begins to clarify as the unwanted material settles to the bottom of the tanks. Moisture is boiled off until the liquid sugar is reduced to a thick, viscous mash, which is pumped into vacuum pans to further concentrate the juice. Nearly dry, the crystals are put into a centrifuge machine, where a residue of molasses is spun off. Again heated to the boiling point, the reliquified sugar is passed repeatedly through charcoal filters. Finally, it is condensed into crystals which are bleached. The resulting product is as pure a chemical as anything you might find in a chemist's laboratory.[5]

In this complicated procedure there are approximately 64 food ingredients eliminated.[6] Potassium, magnesium, calcium, iron, manganese, phosphate, and sulfate are among the discarded minerals; A, D, and the B complex are some of the vitamins destroyed, and essential enzymes, amino acids, fibers, and unsaturated fats are all removed. Most of the B complex vitamins are absorbed into a by-product known as blackstrap molasses, an excellent sweetener whose strong flavor prevents you from using too much of it. All the other important nutrients are in the residue which is processed as feed for livestock, who so far have shown no evidence of diabetes. So-called "raw" sugar is only white sugar to which a little molasses has been added for color and flavor, plus a few minerals, enabling the manufacturer to mendaciously label it "more nutritious."

From 1850 to 1900, world production of sugar increased from one and one-half million tons to 11 million tons. By 1950, production reached 35 million tons, and 20 years later the figure pyramided to 70 million tons. In other words, world sugar production has multiplied fifty times over during the last 125 years. During the same period, the number of diabetics increased on a curve that closely paralleled sugar consumption. In England, when World Wars I and II curtailed sugar use, deaths from diabetes dropped as dramatically as the per capita consumption of sugar, and then rose again abruptly when rationing stopped. In 1907 the Metropolitan Life Insurance Company listed diabetes as the 27th leading cause of death. By 1974 it had been escalated to fifth place.[8]

Noting that heart attacks, arteriosclerosis, and strokes—common causes of death among diabetics—had also become leading causes of death among non-diabetics during that same period from 1907-1974, famed British scientist Dr. T. L. Cleave segregated sugar as the single common causal factor. After years of sifting through the evidence, Dr. Cleave concluded that all these illnesses were actully only different symptoms of the same "sugar disease." According to Dr. Cleave, "Refining of carbohydrates [like white sugar and white flour] produces its very harmful results in three main ways:"[9]

1. Refined sugar is eight times as concentrated as flour, and perhaps eight times as dangerous. The refined product deceives the tongue and the appetite, thus leading to overconsumption. Otherwise why would anyone eat 2½ pounds of sugar beets in one day? Of course, its equivalent in refined sugar is a mere 5 ounces. Overconsumption can lead to diabetes, obesity, and coronary thrombosis, among other dangerous conditions.

2. The removal of natural vegetable fiber can produce tooth decay, gum disease, stomach problems, varicose veins, hemorrhoids, and diverticular disease.

3. The removal of proteins can lead to peptic ulcers.[10]

Dr. Cleave points out that in terms of human evolution, sugar consumption is only seconds old. The human desire for sweets was originally nature's way of tempting us to eat succulent and nourishing fruits and berries. These whole foods, coupled with natural grains, meats and vegetables, enabled humans to thrive and multiply through the ages.

When civilized peoples were able to isolate the sweet-tasting chemical that had encouraged them to eat natural foods packed with vitamins, minerals, and roughage, they disrupted the biological balance that had served them so well for so long. They then went on to combine refined sugar with adulterated, bleached white flour to produce entirely new and totally unnatural foods.

Today, sugar is included in nearly everything we eat. By becoming increasingly more dependent on processed foods, we are ingesting more refined sugar than people did a mere generation ago, when three-quarters of all sugar consumed came from a bag bought at the grocery store. Today, only one-quarter of the sugar consumed is used at the table and for cooking. Besides the obvious processed foods such as cookies, cakes, desserts, and soft drinks, sugar is listed as an ingredient in canned corned beef hash, ketchup, canned and powdered soups, salad dressings, peanut butter, boxed "hamburger-helper" dinners, cheese, TV dinners, luncheon meats, and canned and frozen vegetables. Even sweet fruits are no longer considered palatable unless a thick syrup is added to the can or package. Moreover cigarettes, cigars, and pipe tobacco contain sugar to satisfy this addiction.

Since sugar is relatively inexpensive to produce, the food industry uses it as a filler to replace more expensive ingredients. It is unlikely that people would add as much sugar to the food they prepared as they readily accept in the "convenience" foods they buy. Manufacturers of cake mixes, for example, have found that the addition of copious amounts of sugar adds weight to the box and increases shelf life. *The New York Times* recently reported that "during one four-year period in the 1960s the amount of sugar used in processed foods increased 50 percent."[11]

Roughage is conspicuously absent from the diets of Americans and other industrialized Western peoples. Vegetable, fruit, grain, and bean fibers are necessary for cleaning the teeth and massaging the gums while we eat. Inside the intestines, the fibers expand with water to produce soft, easy-to-pass stools. Moving through the intestines, these vital fibers also act as a natural cleansing agent, removing any bits of debris that might cling to the intestinal walls. Constant ingestion of over-refined foods like sugar and white flour leads to hard, compacted stools that are difficult to evacuate, resulting in constipation. And the problem of body pollution is then further compounded by adding an irritating laxative chemical to the intestinal tract.

Straining to eliminate dense, sluggish waste matter from the bowel can cause colitis (an infection or obstruction of the colon), or diverticulosis (a pouch in the wall of an intestine), which may then become infected. Continued irritation from a highly refined carbohydrate diet may cause either condition to develop into cancerous lesions.

The consequence of indiscriminate sugar use is immediately apparent in the deplorable condition of our teeth. Nearly one in five Americans wears false teeth by the age of 50, and 95 percent are afflicted with tooth decay.[12] In Great Britain, where the annual per capita consumption of sugar is even higher than ours, the problem is worse—more than one-third of all adults over sixteen have major dental work done each year.

Dental caries (decay) were rare in primitive peoples. It was not until the introduction of refined sugar and flour that the problem became alarming. In the few primitive tribes that still exist in various corners of the world, who eat only unrefined carbohydrates rich in nutrients and fibers, dental disease is almost unknown.

Sugar feeds the bacteria normally present in the mouth, and causes them to multiply. These bacteria adhere to the surfaces of the teeth, somewhat like barnacles adhering to the hull of a ship, forming a deposit known as plaque. Although tooth enamel is the strongest material in the body, the bacteria inside the plaque are able to eat through it and attack the dentine inside. A hole is eventually bored through to the pulp that feeds the tooth and the bacteria proceed to the root canal, causing a toothache and eventually destroying the tooth. Meanwhile, the excess sugar inside the stomach has depleted the body's calcium supply, further weakening the tooth's ability to protect itself.

The gums suffer equally, for the soft, sticky nature of over-refined foods does not supply sufficient friction as the food is masticated to keep the gum margins hard, stimulated and tight up around the teeth. Pyorrhea (peridontal disease) is caused when plaque collects at the base of the teeth and forms into a hard deposit called tartar. Rather than affecting the teeth, the bacteria attack the gums and bones that support them. Bleeding and swollen gums are the first warning signal, and unless this difficult-to-cure disease can be checked, extraction of all the teeth will be inevitable. Many dentists say that a good way to prevent tooth decay and gum disease is to eat a lot of chewy foods, such as apples. However, even natural sugar such as that found in fruit will encourage plaque formation (as will all carbohydrates). Brushing, therefore, remains all important.

For years, doctors, medical researchers, and nutritionists believed that an excessive intake of saturated (animal) fats such as milk, butter, and cheese in the diet was a primary contributing factor to heart and artery diseases, the first and sixth leading causes of death in this country. It has long been known that in certain individuals, especially those who are overweight, a fatty deposit known as cholesterol clings to the inside of arterial walls. Over the years, the cholesterol deposits clog the arteries, hardening and narrowing them until the blood cannot pass freely to the heart. Circulation is impaired, the heart receives an insufficient supply of oxygen, and the person tires easily, develops leg cramps while walking, or chest and heart pain. The artery may close off completely, causing a massive heart attack.

Now Dr. Cleave offers startling new evidence that animal fats are not the only culprits; sugars are also. Eggs, cheese, butter, milk, and livestock meats contain fats that were a staple part of the human diet long before debilitating circulatory diseases became an epidemic in western countries. While we are unwisely consuming more of these animal fats than ever before, it is proportionately not nearly so much as the amount of refined carbohydrates. When we fail to manufacture sufficient insulin to adequately convert sugar to glucose the result is a high blood sugar level and an increase in a blood lipid (fat) known as the triglycerides, and there is evidence which suggests that the triglycerides do the greater part of the damage to circulatory tissue.

To illustrate the relationship between coronary disease and sugar, Dr. Cleave cites examples from Africa, where primitive tribes live close by urban, westernized peoples of the same stock. Diabetes and coronary disease is practically unknown to the Zulu tribe, for example, while their cousins who went to the cities to work—and consequently started eating large amounts of sugar—soon began to show the familiar symptoms of deterioration. *"Of one thing the author is very confident,"* Dr. Cleave emphasizes in italics in his important book, *The Sacchar-*

ine Disease, "the key to causation of coronary thrombosis lies in the causation of diabetes (and also of obesity)."

Sugar comes in many disguises, and to effectively combat Sweet Suicide, we must be aware of the potential dangers of its surrogate forms.

Corn Syrup

Derived from cornstarch, corn syrup is essentially a liquid white sugar with nothing of value contained in it. A large portion of the annual intake of sugar is consumed as corn syrup, which is used to sweeten everything from flavored aspirin to wines and cordials. Presented as a pancake syrup, which "has natural goodness and real down-home flavor," this clear, sticky juice is yet another inferior food. The ingredients commonly listed are corn syrup, water, algin derivative, salt, sodium citrate, citric acid, and artificial flavor and coloring.

Glucose

Also made from cornstarch, glucose is a leading contributor to the adulteration of food. Since there is no law requiring that glucose be listed on a product label, the food industry indiscriminately mixes liberal amounts of this cheap filler in jams and jellies, preserves, processed cheeses, candy, canned milk, store-baked goods, and soft drinks. Disreputable manufacturers saturate dried fruits with this sugar to add weight, charging more for the nutritionally valueless extra substance.

Because it is not as sweet as white sugar, glucose is all the more dangerous. People who buy the products listed above may be consuming huge quantities of glucose without realizing it, because they can't taste it. As high in calories as sugar, glucose is in effect a predigested food that undergoes no processing in the stomach or intestines. Back in the 1920s, Dr. Harvey W. Wiley warned of how this pure energy source can interfere with natural digestion. He told how the flooding of our stomach with dextrose (glucose) would create a situation which would require half a dozen artificial pancreases for our body to be able to cope with it.

Waging a battle to ban glucose from food, Dr. Wiley managed to mobilize a sizeable section of the American public, and a Congressional hearing was conducted. The sugar refining interests, in turn, mobilized their considerable resources and brought pressure to bear on the Department of Agriculture, Congress, and the President. The protest was hastily dismissed.[13] The battle had been lost, but the consumer was at least beginning to realize he was being misused by the greedy manufacturers he had enriched.

Maple Syrup

Can anything that flows from the living core of a tree be bad for you? The answer is yes and no. Yes, maple syrup does contain nutrients, but not enough to make it into a healthy food. As with any sugar, there is a tendency to use too much of it, a temptation increased by maple syrup's good flavor.

Even taken in moderation, maple syrup can still be bad for you, for most of it is no longer an uncontaminated, natural product. In 1962, the F.D.A. sanctioned the use of a paraformaldehyde pellet to increase the output of maple sap. Plugged into the tap hole, the pellet destroys bacteria normally found there, but which when eliminated can no longer slow down the flow of sap during the beginning of the growing season. During this period, the tree's internal chemistry is producing growth substances which mix with the sap, giving it a "buddy" taste. To overcome this, a culture of bacteria is added to ferment the drawn syrup and remove the unwanted flavor.

The F.D.A. has established a level of tolerance for the long-lasting and slow-dissolving paraformaldehyde pellets, meaning that a residue of this chemical is now an integral ingredient of most American maple syrup. However, sugarbush farmers in Vermont refuse to use the pellet and the Canadian Food and Drug Directorate has wisely banned its use throughout Canada.

Honey

Throughout most of man's history the basic sweetening agent has been honey. Sweeter than sugar, honey is a preferred substitute for white sugar and other sweeteners, as less is needed to provide the desired taste. Hence, fewer calories are used. Honey contains about 70 to 80 percent simple sugars, and the rest is composed of minerals, vitamins, and important trace substances. The darker varieties, especially those left in the comb, contain significantly higher amounts of these life-giving elements.

Pure honey is slightly cloudy with a residue at the bottom of the jar. Most consumers are not aware of this fact, assuming that the cloudiness and residue are impurities. Because of this, many manufacturers heat and filter the honey, destroying some of the vitamins. The clear, brilliant and easy flowing honey that many consumers prefer has lost much of its delicious flavor and nutritional value through commercial processing.

Bees are highly sensitive to pesticides and they avoid gathering nectar from polluted fields. If they accidently happen onto one, they ususally die before returning to the hive. Honey is, therefore, one of the few foodstuffs widely available that has virtually no pesticide residues (sugar cane and sugar beets are both sprayed and artificially fertilized), a fact that should be reassuring but is misleading since other chemical residues are introduced which are essentially as harmful.

Before honey production became big business, bee-keepers used a harmless bee-escape device to separate the insects from the honey saturated combs, and brushed off remaining strays with a soft brush. But with the drive for big profits and increased production came the inevitable abuses, and now irritating chemicals are used to do the job more efficiently. Poisonous sprays, including phenol (carbolic acid), benzaldehyde, and nitrous oxide remove the bees, but they may also find their way into the honey. Often, empty combs are fumigated with moth balls (paradichlorobenzene) or methyl bromide, whose residues contaminate the honey as it is produced.

Although honey is nutritionally preferred over white sugar, that does not suggest it is safe to consume it in large quantities. It can also contribute to tooth decay, obesity, diabetes, and some of the other diseases sugar addicts are heir to.

Cyclamates

In the 1950s, millions of Americans living the good life in the burgeoning post-war economy decided they were getting too fat. The fast-rising rates of death from coronary and cardio-vascular diseases, and cancer of the colon concerned them, too, but not enough to give up sweets. But sugar made them obese and therefore unattractive, creating a conflict between what they wanted to eat and how they wanted to look.

The food industry satisfied both the conflict and the craving by creating products that contained artificial chemicals that made foods taste sweet without adding calories. About 30 times sweeter than sugar, cyclamates had been synthesized from coal-tar derivatives in 1937, and manufacturers touted it as the perfect solution for dieters—a sweet food with no extra calories, not to mention no nutritive value. Unlike the older artificial sweetener, saccharin, cyclamates produced no bitter aftertaste, and they had the added advantage of being usable in cooking.

Originally cyclamates were approved by the F.D.A. for use primarily by diabetics, but the ruling was loosely worded. The food industry subtly changed the warnings on the labels required by the F.D.A. to make it seem as if their products could be safely consumed by anyone. Thus, "for those who must restrict their intake of sugar" became "by persons who desire . . ."; "should only be used by . . ." was twisted to "recommended for . . ." and "non-nutritive" became "no calories."

Cyclamate manufacturers aimed their major advertising thrust at people who wanted to reduce their weight. Cyclamates were introduced into jellies, jams, preserves, ice cream, soda pop, fruit juices, candies, cookies, breakfast cereals, "900 calorie" meals, puddings, syrups, canned soups, salad dressings, and bread. By 1969 an estimated 175 million Americans wer consuming 20 million pounds of cyclamates per year.[1]

Faced with this flood of uses, in 1962 the F.D.A. requested the Food Protection Committee, Food and Nutrition Board (The National Academy of Sciences—National Research Council) to review the safety of these artificial sweeteners. After several admissions of a lack of definite knowledge, the Committee concluded cautiously: "There is no evidence that the use of non-nutritive sweeteners, saccharine and cyclamate, for special dietary purposes is hazardous." However, the conclusion did not address itself to the primary question, since cyclamates were being used widely and indiscriminately, and not only for "special dietary purposes."

In the next few years, medical reports containing specific evidence of the harm done by cyclamates began to appear. Damage to fetuses, diarrhea, inhibited growth, damage to kidneys, liver, the intestinal tract, thyroid and adrenal glands, changes in blood-coagulation, and blocking action of antibiotics were some of the adverse reactions to cyclamates. Nevertheless, the F.D.A. extended the use of the synthetic sweetener to new groups of foods.

Finally, in late 1967 the N.A.S.-N.R.C. again reviewed the safety of cyclamates and, after examining reports of adverse reactions received throughout the world, they reversed their original conclusion and recommended that, "totally unrestricted use of the cyclamates is not warranted at this time." Meanwhile, the F.D.A. was conducting tests of its own on laboratory animals. They could see for themselves that rats fed heavy and continued doses of cyclamates developed bladder cancer and suffered chromosome damage, clear evidence that such radical changes could also happen in humans.

Acting under the provisions of the Delaney Amendment, which makes it mandatory to ban any food ingredient if it has been shown to cause cancer in animals or humans, Secretary of Health, Education and Welfare Robert H. Finch imposed a total ban on cyclamates in October, 1969. Finally, 19 years after the introduction of cyclamates, and approximately 14 years after their harmfulness was first revealed, this poison was taken off the market.

Saccharin

With the ban of cyclamates, saccharin was the only low-calorie sugar substitute on the market. As the demand for a non-fattening sweetener was greater than ever, food manufacturers simply switched to this equally inexpensive sugar substitute and added it to the same foods. Saccharin, another coal-tar derivative 300 times as sweet as sugar, was discovered in 1879 and has been in use since then. Its chief drawback is an astringent, metallic after-taste, making it inferior to cyclamates.

Since saccharin is so intensely sweet, it is less bulky to handle than cyclamates and cheaper to produce. A pill the size of a large pinhead is enough to sweeten a cup of coffee. Anxious to capitalize on these appealing properties, the industry found a way to mask the aftertaste by adding approximately 0.2 percent of glycine, an amino acid. Now, as a result of this simple addition the formerly heavy intake of cyclamates has been surpassed by the current consumption of saccharin. The F.D.A. has only recently begun to seriously investigate a reported link between cancer and saccharin. In March, 1970, researchers at the University of Wisconsin reported that 50 percent of the mice who had saccharin pellets injected into their bladders developed cancer. Although this is not the manner in which artificial sweeteners are consumed normally by humans, the results of the experiment reveal more than just a casual link, and more thorough studies should immediately be undertaken. An examination conducted for the F.D.A. by The National Academy of Sciences-National Research Council only produced the following evasive statement: "Our studies tend to confirm the evidence from animal studies that saccharin has a wide range of safety for acute and short-term exposure."

Based upon newly discovered evidence, the F.D.A. banned the use of saccharin on March 9, 1977, an action to take effect in July, 1977.

The Sugar-Coated Teat

The craving for sweets usually begins shortly after birth, when a bottle of sugar-laden cow's milk is given to the infant. When a baby cries often, he or she is given a sugar-coated pacifier. At two or three months, the infant is given some processed solid baby food, which also contains sugar. From baby foods the child progresses to sweetened breakfast cereals, with candy and cake frequently introduced as rewards for good behavior.

The fat baby that everybody loved so much and mistakenly thought was so healthy, grows up to become a fat adult who is permanently addicted to sugar. Overfed during the crucial growing years, a person develops an excess of fat cells in his or her body, which permanently require satiation. Whenever the fat level drops in a significant number of such cells, they require another dose from the liver, and hunger results. We all know at least one fat person who is constantly dieting to bring his or her weight down. It is not always a matter of having inadequate willpower, often it is a biological malfunction directly resulting from a long-term excessive use of sugar, a malfunction most Americans suffer to varying degrees.

The only solution is to avoid *all* refined white sugar and bleached white flour in whatever forms they are disguised. Artificial sweeteners are not a viable alternative, for they, too, cause degenerative disease. Moreover, they keep alive a "sweet tooth," and it is only a matter of time before the user returns to sugar. Choose desserts from the abundant, fragrant, juicy fruits of the earth. Pastries and cakes made from whole grain flours will reawaken tired tastebuds, and sugar will eventually taste like the refined carbohydrate it is.

Notes

1. U.S. Department of Agriculture, Economic Research Service, 1975.
2. Martin, W.C., "When Is Food a Food and When a Poison?," *Michigan Organic News*, March, 1957.
3. Himsworth, H.P., *Clinical Science*, 1935.
4. Banting, F.G., *Strength and Health*, May-June, 1972.
5. Strong, L.A.G., *The Story of Sugar*, George Weldenfeld and Nicolson, London, 1954.
6. *Ibid.*
7. Viton and Pignalosa, *Trends and forces of world sugar consumption*, F.A.O. Commodity Bulletin Series, No. 32, United Nations, New York.
8. Cleave, T.L. *The Saccharine Disease*, Keats Publishing Co., New Canaan, Conn., 1975.
9. *Ibid.*
10. Dufty, William, *Sugar Blues*, Chilton Book Co., Radnor, Pennsylvania, 1975.
11. Lyons, Richard D., "Sugar in Almost Everything You Eat," *The New York Times, News of the Week in Review*, March 11, 1973.
12. Consumer Reports, *The Medicine Show*, 1974.
13. Wiley, Harvey W., *The History of a Crime Against the Food Law*, published by the author in Washington, D. C., 1929.
14. Trager, James, *The Food Book*, Avon Books, New York, 1972.

The Meat Factory—
Drugged, Dirty, and Deadly

Concerning the meat available at your supermarket, so many chemicals are ingested by and injected into the 238 pounds of beef, pork, and poultry that each American consumes annually[1] that it is impossible to say how much is safe and how much is potentially harmful. As recently as a generation ago, most food animals lived out of doors and fattened themselves on grasses and whole grains to produce edible flesh that was pure and fairly safe. Unfortunately, the family farm has been replaced by corporation-owned "animal factories" where livestock and poultry are forced to live in crowded environments that are frequently unclean and germ-ridden. The animals are kept alive and fattened by the continuous administration of tranquilizers, hormones, antibiotics, and 2,700 other F.D.A.-approved drugs. The process starts even *before* birth and continues long *after* death.

Beef—The Flesh Is Weak

On a warm day when the wind is right, you can smell a cattle feedlot from several miles away. Hundreds of thousands of animals are packed hoof-to-hoof in a honeycomb of small pens, where they stand shank-deep in their own excrement, eating and defecating until it is their turn in the slaughterhouse. The feedlot is the beginning of four months of inhumane treatment that will make their violent deaths a blessing by comparison.

When they are a year old and weigh approximately 500 pounds, the cattle are herded from the spacious rangeland where they grew up, into trucks and boxcars. After a journey, which is often long, to a commercial feedlot located near the slaughterhouse, the panicky, hungry animals are prodded into a tank filled with pesticides to rid them of parasites and flies, then they are packed into pens, 400 to an acre. During the next four months the animals are fed nothing but starchy low protein grains which constantly fill the long feed bunks, and get fat. To reduce the time this process takes naturally, bright sodium-arc lamps are turned on at night to encourage the confused animals to feed 24 hours a day. By combining overfeeding with over-stimulation, cattle can now be slaughtered two months earlier than they were 20 years ago.

The steers each eat nearly 30 pounds of grain a day, which they convert into about three pounds of muscle and fat. But this is not fast enough for the meat industry; they often shoot pellets filled with diethylstilbestrol (DES), a female hormone, into the animal's ears to make them grow fatter on less feed. Another female hormone, melengestrol acetate (MGA) eliminates the sex drive of cattle, who might otherwise lose valuable pounds by engaging in sexual activity. In fact the breeding of cattle is, for the most part, completely manipulated by man by means of artificial insemination.

Crowded together in such unnatural, unsanitary conditions, cattle are apt to become restless and violent, so tranquilizers are added to their feed to calm them down. The animals stand nearly motionless in their feedlots, calmly munching grains all day and night, their eyes glazed over as if they were hypnotized. Actually, they are in a kind of stupor from the many drugs administered to them.

By being forced to eat such huge amounts of rich grains, however, the animals' bodies rebel and new problems are created. A large percentage of feedlot cattle develop painful liver abscesses, which slow down their weight gain. This malady is combated by introducing into their feed yet another chemical, an antibiotic such as oxytetracycline. To control epidemic diseases like diarrhea, respiratory ailments, and foot rot, the cattle are injected with a stronger antibiotic such as streptomycin. The drugs have the added advantage of making cattle grow faster by destroying bacteria in the intestinal tract that normally inhibit the metabolism involved in the animals' conversion of grain into protein.

After four months of being converted into chemical food machines, the cattle are bulging and overloaded with fat. On the way to the slaughterhouse they are given one last chemical treatment; an injection of papaya enzymes to break down muscle fibers, so the meat will be tender by the time it reaches your kitchen. The dose is carefully measured, for too much will cause the animal to die of convulsions on the spot. Inside the slaughterhouse, cattle are secured to conveyor belts and stunned into unconsciousness electrically or with captive-bolt pistols. The animals may not actually be dead, but it hardly matters since they are bled and butchered into their component parts within an hour, often while their hearts are still beating!

It is not a pretty picture, and it explains why more and more people are choosing to limit meat in their diets. But meat-eaters' worries don't end with humane considerations, or with hormones, antibiotics, and tranquilizers; the F.D.A. permits residues of these potentially dangerous drugs to exist in our primary source of protein. At the supermarket, or the wholesale meat distributor, for instance, steaks, chops, and roasts are sometimes dipped into antibiotic solutions or chemical preservatives to increase their shelf lives. We cannot protect ourselves against these additives because there is no law requiring that they be listed on labels.

Even the birth of food animals is promoted by injections of synthetized hormones (estrogen and progesterone) to increase the fertility of cows. At present, new hormones are being formulated to induce the birth of twins rather than the usual single calf, an obvious economic advantage if this could be achieved in every animal. Contact between bulls and cows is not as frequent as it once was on most breeding farms, especially where artificial insemination is standard practice. Obtained by stimulating bulls with dummy cows, the semen is treated with the preservative glycerol and frozen until it is needed. Today, a single bull can be used to impregnate as many as 20,000 cows. Close to half of the dairy cows and cattle born in this country annually are sired this way.[2]

The surplus bulls are castrated by the implantation of female hormones, which turns them into lazy, fat steers which, in the public's mind, is the best beef. Actually, the meat is too fatty, and the animal often develops illnesses that reduce the quality of its beef yield.

Another technique of artificial breeding involves the insertion of a hormone-saturated sponge into a cow's vagina near the womb to prevent her from ovulating, so that she can be impregnated at the same time as hundreds of other cows at the convenience of the rancher. The sponge is removed with a drawstring about three weeks later, and the animal is shot full of opposing hormones to induce "super-ovulation" before being serviced. By using massive doses of the same hormones contained in human birth control and fertility pills which can be hazardous to human health even in small quantities, the cattle industry keeps their stock constantly pregnant, and exhausted. The many deformed calves that are born are killed at birth and ground up into pet foods or thrown away.

Those calves not destined for feedlots are taken away from their mothers prematurely and fed only milk or a milk substitute to produce the familiar white flesh of "milk-fed" veal. The cost to the animal for this misguided consumer luxury is a short, mal-nourished lifetime of weakness and anemia and its meat is deficient in a number of vitamins and minerals.

This chemical tampering with food animals is doubly alarming because of the biological similarity between them and humans, the animals who eventually consume the treated flesh. Despite assurances from the F.D.A. and the meat and the drug industries that the use of drugs in the production of livestock poses no health problems, evidence exists to the contrary. "We cannot wait for decades of hazardous exposure to continue before evidence in man is obtained," warns Umberto Saffioti, a cancer researcher at the National Institute of Health.[3] Dr. E. Boyland, another leading cancer researcher, states that "not more than 5 percent of human cancer is due to viruses and less than 5 percent to radiation. Some 90 percent of cancer in man is therefore due to chemicals."[4]

Anyone who makes the claim that chemical contaminants foreign to animal tissue are benign or beneficial is simply fooling himself. Research in this area is still shockingly inadequate, and the chemicals in question have not been in use long enough to determine their long-term effects on the mass of people who are unknowingly consuming them. Some cancer takes from 10 to 25 years to develop from the first irritation, and animal drugs were introduced approximately 25 years ago, so it is too early to know for certain the effects of most of these drugs on humans.

Hormones—Biological Uncertainties

Since the reproductive apparatus in the human is quite similar to that of the cow, women who eat meat daily run the risk of upsetting the hormone balance of their bodies with residues of DES and MGA, just as cows do. Consumer advocate Daniel Zwerdling claims that the F.D.A. has secret evidence that MGA "inhibits reproduction," which the agency refuses to disclose because a major drug company is involved. Nutrition Institute of America (N.I.A.) researchers were unable to confirm this. However, many scientists are of the belief that trace residues of the hormones used in cattle could cause difficulty with some women's menstural cycle. Zwerdling goes on to quote an anonymous F.D.A. veterinarian who agrees with these other scientists that "trace residues of the hormone could disrupt a woman's mentrual cycle."[5]

It has already been established that miniscule amounts of DES can produce excessive menstrual bleeding in women, and impotence and a low sperm count in men. According to *The Medical Officer,* a British government-sponsored health magazine, hormones contained in chemically fattened meat have caused English girls to reach puberty three years earlier than a generation ago.[6] DES contamination has also been implicated in breast cancer, fibroid tumors, and leukemia. Farmers who inhaled DES dust while mixing it with feed soon realized the drug's potency when they developed atrophied testicles, sterility, and tone changes in their voices.

In 1971, a positive link between DES and a mutation-cancer was finally proven; women who had taken this synthetic hormone 20 years earlier as a fertility drug produced daughters susceptible to a rare but fatal form of cervical cancer. Faced with proof, the F.D.A. had no choice but to impose a ban the following year, an action that seemed like a victory for the people. Unwilling to give up $90 million worth of business so easily, the drug industry fought back in court, basing their case on the specious argument that the level of residues in meat did not present the same dangers as taking DES directly. Since research was incomplete, and the F.D.A. and the Department of Agriculture had neglected to develop testing methods that could determine how much DES residue was contained in meat, the court rescinded the ban because of lack of evidence. Since then the F.D.A. and U.S.D.A. have developed chemicals and gas spectroscopes that can detect the tiniest amounts of chemicals in foods, and the residues of DES have been found to be on the *increase*.

In 1976, F.D.A. Commissioner Dr. Alexander M. Schmidt blamed the agency's failure to protect the public from this proven carcinogen on drug industry lawyers: "When the lawyers go to work, they often remove the action from the scientists—from the people who can best make benefit/risk decisions. We have again proposed a ban of DES as an animal-feed additive."[7] Yet, even if the F.D.A. succeeds this time, it probably will make no difference to our health, for the agency has already approved half a dozen potentially dangerous animal drugs that can be substituted for DES. "We'll be going to other hormones soon," says the Washington director of the American National Cattlemen's Association confidently.

All estrogens are suspected carcinogens, including some of the new combinations of several pre-existing sex hormones. The expensive public relations firms that drug manufacturers hire to influence public opinion in their favor offer the low incidence of cattle deaths as proof that hormone-treated feed is not harmful to mammalian tissue. The argument is just another example of industry obfuscation, because the animals—unlike humans—do not live long enough to develop fatal tumors. The argument that sex hormones are harmless because they already exist in the body is not valid since synthetic hormones are not the same as those existing in nature.

Antibiotics—Increasing the Risk

Just as some species of insects have developed a resistance to pesticides, an increasing number of bacteria have become immune to antibiotic drugs that once destroyed them. This is partly due to the widespread dosing of farm animals, who receive half of the nation's annual antibiotic production. Inexplicably, a rancher can purchase penicillin by the pound at a feed store, while requiring a prescription for as little as one tablet for his own use. With antibiotics so freely available, it is not surprising that some ranchers use many times the recommended amounts in order to coax maximum growth from their animals.

When antibiotics are regularly ingested with food, many people become allergic to them, and the drugs are transformed into killers rather than saviors. It has been estimated that as many as 20 million Americans break out in rashes when injected with penicillin, and that hundreds go into shock and die. Others suffer stomach upsets because the drugs kill all bacteria in the digestive system— the beneficial ones that help digest food and fight disease, as well as the unhealthy ones. The greatest danger for most of us, however, is that highly toxic bacteria build up an immunity to the only drugs able to kill them. "One of the biggest problems in the world today," says Dr. David

Smith, an expert on infectious disease and an advisor to the F.D.A., "is bacterial deaths due to antibiotic-resistant bacteria. We may eat a hamburger with antibiotics in it now, and not feel the results until we're 60. Then maybe we'll get a heart valve infection; they'll give us some penicillin and discover it's useless."

The massive use of antibiotics in livestock was seriously questioned in 1971 by Dr. H. Dwight Mercer, the deputy director of the F.D.A.'s veterinary research farm. After raising livestock for a year and a half with and without antibiotics in clean conditions on uncontaminated feeds, Dr. Mercer concluded that the drugs made no difference to growth. The antibiotics helped only as the pens got filthier and the quality of food worsened. "The worse the conditions, the better the response," he reported. Other studies have borne out Dr. Mercer's findings: antibiotics work best in the presence of manure-encrusted filth, disease, crowding, substandard food, extreme cold or heat, and extreme stress.

"We use too many drugs in place of good management and sanitary practices," says Dr. William Buck of Iowa State University. "We could put our emphasis on a little better management and disease control instead of drugs, and do just about as well. Of course, that would put some hardship on the farmer; it would require more physical labor to clean out the pens, and keep them washed better."

Grass-Fed Beef and World Hunger

Leaner than the usual grain-fed variety, and lower in cholesterol, beef fed on grass have been presented as a new solution to a few American health problems, and an aid to world hunger. In addition, grass-fed beef is cheaper to produce, and presumably the savings would be passed on to the consumer. Although it seems like something new, feeding beef on grass in the open range was standard practice until after World War II. It was then that great surpluses of corn and other grains began to accumulate, which were partially disposed of by feeding them to cattle. The result was the familiar, tender beef, well-marbled with thick white fat, and a possible contributing factor to the increase in the heart disease rate in this country.[8]

Because grass-fed beef is a darker red with less fat marbling, it has not met with wide consumer acceptance, even though it is much healthier than the regular supermarket variety. Its flavor is as good, if not better, but more important the millions of tons of valuable grains slated for cattle feed could be distributed to starving peoples around the world. Being avid meat-eaters, Americans consume the equivalent of a ton of grain a year on the average, all but 150 pounds of it in the form of meat, eggs, and dairy products from animals that have been fed on the grain. This is five times the amount available per person in countries like India, where the grain is eaten directly. In a very real sense, we are gorging our food animals with extra calories that they don't need, and making them eat for two; themselves and the ultimate consumer, us.

In fact, even if we substituted grass-fed beef for our annual consumption of meat, we would still be eating far more than is good for us. An optimum diet is one that draws on a variety of foodstuffs, where the daily protein allowance is from a combination of sources, including grains, vegetables, cheese, eggs, poultry, and fish. A diet overbalanced with a high-nutrition food like beef will crowd out other vital nutrients.

Among the more common health problems that can result from eating too much beef are bowel problems and obesity. There is little roughage in beef and most other meats, and poor bowel movements result from a lack of bulk needed to move food through the intestines at a healthy rate of speed. Such foods as lettuce, carrots, cabbage, and unprocessed grains must be consumed daily to prevent such diseases as diverticulosis, the formation of many tiny sacs on the sides of the intestines. The National Cancer Institute recently released a five-year study that suggested a link between a high consumption of beef and cancer of the colon.

As beef is a major source of calories—eight ounces of round steak contains 500 calories—it can also make you fat. And needless to say, heart disease is such a rampant problem that public health agencies actively urge each of us to cut down on all animal fats, especially well-marbled beefsteak. Since grain-fed beef is fraught with the attendant dangers from the chemicals and hormones cattle are treated with, protect your body by eating more nutritious, safer proteins.

The Poultry Problem

Like most people, you probably believe that chickens are raised on a cozy little farm, where they run free in the yard pecking at feed thrown to them by the farmer. This idealized picture is far from the truth. In reality, chickens grow up on computerized farms in a factory environment that is about as far from nature as you can get.

Chickens that used to run free to scratch and root in the soil for the valuable nutrients they found in earthworms, grubs, and other larvae are as outmoded as the horse and buggy. Today's bird never sees sunlight or breathes fresh air, since it must spend its short life in a 12″ by 12″ cage with four other birds in huge windowless, temperature-controlled sheds. Its day is completely programmed by electronic computers that select its feed and sends it to block-long troughs on conveyor belts. At specified times, powerful sprays of water shoot through the pens to wash away excrement, sending the birds into a flurry of feathers and outraged clucking. Bright fluorescent lamps are left on 16 hours at a time to encourage feeding, and the chicken will probably never come in contact with a human being until it is eaten.

In this way, poultry farmers have been able to boost production to 12 billion pounds of poultry meat a year[9]—a 600 percent increase since World War II. On a modern American chicken farm, one person can take care of 30,000 birds at a time by keeping the birds in confinement and doctoring their feed with a variety of drugs and hormones. As with cattle, chemical treatment speeds up their growth and gets them to market sooner. In 1945 it took 16 to 17 weeks, with fifteen pounds of feed, to produce a three pound chicken; today, it takes only nine weeks and half as much food. By speeding up the birds' metabolism, one poultry farmer can produce more than 160,000 birds a year, a highly profitable enterprise.

While mass-production has benefited the consumer by lowering the price of chicken from 60¢ a pound in 1950 to an average of 49¢ a pound in July, 1976, it has not been without cost both to bird and consumer. Being crowded together, the birds are subject to great stress that cannot be relieved by scratching through the barnyard. Consequently, they may peck and cannibalize each other, a psychotic behavior that seriously threatens productivity and profits. One way this reaction is controlled is by bathing chicken houses in red light so that the birds will not be able to see the anger-provoking red combs of their cell mates. If this method fails to work, the upper halves of the birds' beaks are snipped off so that no matter how insanely they peck, little damage is done. And often, tranquilizers are mixed with the feed to help calm the chickens.

The most apparent result of speeding up poultry growth is its flat, tasteless flesh. When chickens were killed at four or six months of age, they had time to develop flavor, but today we have no choice but to buy mass-produced, underage poultry that is universally bland. The industry is aware of this problem, but instead of improving the breeding and growth process, they have come up with ways to mask inferior quality. Chemists have invented artificial dyes that are added to feed to transform the pasty whiteness of chicken flesh into a rich, golden color. Those of you who have had the good fortune to purchase chickens raised in a natural environment will realize that these dyed birds look unreal; the skin of a healthy chicken is a *rosy* yellow that no chemical can duplicate.

Regardless of color, the flavor of mass-produced poultry is a pallid echo of what it once was, when families looked forward to "spring chickens" culled from the spring hatch of laying flocks. To give the chickens flavor, some farmers inject the birds prior to slaughter with the enzyme hyaluronidase and a seasoning mixture of garlic, sage, nutmeg, and thyme. The enzyme helps metabolize and disperse the herbs throughout the bird, and their fragrant aroma will later deodorize the acrid scent the enzyme gives off during cooking.[10] Currently, U.S. Department of Agriculture researchers are attempting to cross chickens with quail to solve the flavor problem.

During their brief lifespan, chickens are particularly vulnerable to certain diseases because of their close, cramped quarters. The most rampant of these is coccidiosis, a parasitic condition that is on the increase despite new methods of control. As the U.S.D.A. explains it, "Farm-raised flocks generally suffer less from coccidiosis than flocks raised under crowded conditions. Because of the space over which farm birds can range, their chances of picking up heavy or fatal doses of *coccidia* are less great ... This tendency to crowd birds sometimes has been carried to the point where control measures for parasites and other disorders, particularly respiratory diseases, are inadequate.

"Crowding should be avoided, because it favors disease by contributing to a build-up of the number of disease organisms. It also deprives the birds of the exercise they need. A comfortable, sunny, well-ventilated poultry house, free from drafts or dampness, favors good health."[11]

The U.S.D.A. recommends that infected birds be isolated, a method considered impractical by farmers because of the manpower needed. Individual medication is also rejected for the same reason, and building up flock immunity is not even considered. It is easier to include a mixture of drugs in chicken feed to kill all bacteria, beneficial or harmful. Nitrofurans made from poisonous arsenic, and antibiotics are part of the birds' diet from birth, and, as with cattle, these drugs have the added advantage of speeding up growth. Arsenic in the form of arsanilic acid has been fed to poultry since 1950 to "make your hens work harder," according to the advertising claims of one manufacturer. They are said to cut feed costs by one-third by making the birds develop faster; to impart a healthier color (augmented by dyes); and—most important—to yield more profits. While these poisons no doubt increase the farmers' efficiency, they also introduce another harmful chemical into man's system.

It is an F.D.A. rule that nitrofurans must be discontinued in poultry feed for a minimum of five days before slaughter so that their livers—where arsenic collects—can have time to cleanse themselves. Most poultry raisers fail to adhere to this rule. And since government inspection is notoriously lax, you run the risk of arsenic poisoning every time you consume supermarket chicken livers.

Antibiotic residues present the same threats to health in poultry and beef, although being leaner, chicken is the lesser of two evils. While they foster rapid growth, the real reason antibiotics are used is because of the filthy conditions that exist on many chicken farms. By applying mass-production methods that may work well on inanimate objects to living organisms, chicken farmers have traded vigor and disease-resistance for an often sickly, processed bird. It no longer matters if the birds are poorly bred, for arsenic and antibiotics will keep them alive and fat until they reach premature maturity. In the meantime, the birds may develop resistant funguses and mold infections in their intestinal tracts from prolonged ingestion of antibiotic feed supplements, and they may pass them along to man as food poisoning.

Some of us have had food poisoning from time to time without realizing it. Contrary to popular belief, it is not ptomaine but the salmonella bacteria that causes stomach cramps, severe diarrhea, and vomiting. A milder cousin of typhoid fever, salmonella is often mistaken to be intestinal flu, the 24-hour virus, or summer flu and the doctor simply puts the patient to bed for a few days with a stomach medication, knowing that he or she will recover from the unknown but familiar malady within a few days. This infectious organism is common in meat—especially chicken—eggs and egg-based products, and improperly stored food, and reaches its annual peak during the warm summer months. Salmonella poisoning is one of our most widespread undiagnosed diseases.

The New Scientist pointed out the problem in 1967 when the magazine reported: "There is now a mass of evidence showing that the misuse of antibiotics as growth-promoting food supplements and as mass prophylactic agents has caused a serious increase in bacterial drug resistance in recent years. The threat to human health and animal health has been made abundantly clear, and warnings from experts in the field have mounted over the past year."[12] What the report means is that should some of us catch a particularly virulent strain of salmonella, it is questionable whether penicillin would be able to check the antibiotic-resistant bacteria.

Public health officials believe that the full extent of the illness is not known because salmonella poisoning is not easily diagnosed. However, estimates of incidence range as high as 38 million cases a year in the United States alone.[13] In two of the most recent mini-epidemics, 107 people became ill in Spokane, Washington in July, 1966, after eating barbecued chicken, and two died; on June 6, 1970, 700 people who attended a picnic in Columbia, South Carolina were afflicted with salmonella gastroen-teritis, and many required hospitalization. To protect yourself from this infectious disease—which does not make itself known through off-odors or tastes—you must remember to wash your hands before and after handling raw chicken (and other meats), and to cook it thoroughly since high temperatures kill the organisms.

The year of the *New Scientist* report—1967—the U.S.D.A. initiated an unscheduled national inspection of poultry processing plants, and uncovered the fact that the poultrymen were scrimping on sanitation at the expense of the public's health. One of every five birds was judged to be unfit for human consumption and 50.8 percent were contaminated with salmonella bacteria. Many of the birds

inspected showed "gross lesions of disease," and "contamination of the body cavity with stomach contents of fecal
materials." As a result, the Wholesome Poultry Inspection
Act was passed in 1968, and processing plants were
ordered to clean up their facilities or face heavy fines. For
a time, poultry producers complied with the stringent new
regulations, but with inspections infrequent and enforcement almost non-existent, conditions are essentially the
same as they were before.[14]

In fact, the U.S.D.A. later lowered standards instead of
raising them, in one major respect. Leukosis is a viral
cancer peculiar to chickens, and it is the largest single killer
of domesticated fowl, costing the industry millions yearly.
The Surgeon General estimates that over 90 percent of
chickens from most flocks in this country and abroad are
infected with leukosis viruses, even though a much smaller
percentage develop overt neoplasms (tumors)."[15] To help
the industry over what it considered an "economic
hardship," the U.S.D.A. relaxed its ruling on condemned
birds in 1970. A panel appointed by the agricultural
agency suggested that chickens infected with cancer virus
be permitted on the market "if they do not look too
repugnant," and that tumors should be simply cut off
while the rest of the chicken could be sold as chicken parts.

Chicken meat could also be contaminated with
cancer-causing, growth-stimulating DES hormones, as are
beef and pork. In the late 1950s, a New York restaurant
worker attained immortality in medical textbooks as an
example of gynecomastia—the growth of full-sized
female breasts on a man. At the time, the poultry industry
was producing a tender-meated rooster called a capon-
ette, whose succulent flesh was not a result of
castration—as in the capon—but the implantation of a

pellet of DES. This synthetic chemical that mimics the action of natural female sex hormones was implanted in the chickens' necks unbeknownst to the kitchen worker. After serving customers the wings, legs, and breasts, the employee cooked the leftover chicken necks for his nightly dinner, and over a period of a year he ingested a countless number of DES pellets and developed startling anatomical abnormalities. Admittedly, the average citizen would not be exposed to such excessive amounts of DES, but the F.D.A. wisely banned the implantation of DES pellets in poultry (but not in beef). Despite the ban, however, hormones are still a part of our chicken dinners, for DES is still permitted in poultry feed and their drinking water.

By the time the chicken reaches the consumer, it could contain an incredible amount of bacterial contaminants and chemical additives, plus copious amounts of water to make it weigh more. To counteract the unsanitary conditions that exist on some assembly lines, poultry farmers dip the bird in a final bath of yet another antibiotic, and soak it in near-freezing water that leeches out what is left of the flavor. Foreigners who eat poultry in this country often complain about its lack of taste, which is not surprising; most European countries forbid the importation of American chickens on not only esthetic but also health grounds. They prefer to let their chickens roam free in the sunlight, the way nature intended.

The Not-So-Good Egg

By rights, the egg—not milk—should be regarded as nature's most nearly perfect food. With the exception of vitamin C, the egg is a balanced source of all the important vitamins and minerals necessary to maintaining good health, and it contains about six grams of high-quality protein. Although the egg has been accused of being a major contributor to artery-clogging cholesterol, this is only partially true. More cholesterol is created from animal fats in the form of meat, milk, butter, fatty cheeses, and sugar, stress, drinking, and smoking. Like all highly nutritious foods, the egg is also highly susceptible to contamination.

Unfortunately, eggs are vehicles for the residue of a variety of additives present in the diet and environment of laying hens. Pesticides are bound to be present from the contaminated grains frequently used as feed, which is also treated with antibiotics, yolk "improvers," and shell hardeners. The laying hen is a little better off than her broiler and roaster kin, for she is housed in well-ventilated, individual pens that are kept relatively clean. A flock animal by nature, the laying hen is denied her place in the barnyard, the "pecking order" of the birds' caste system, her turn to mate with the rooster, and the abundant herbage and insect life found out-of-doors. All these denials add up to producing eggs that do not taste as good; the yolks are thin and break easily, and the egg has a soupy consistency which simply does not match the free-running, healthy-chicken-laid eggs.

The treated food consumed by poultry factory hens year-round produces eggs with yolks that are pale and watery, necessitating the addition of coloring matter to the feed to darken them. Dr. Franklin Bicknell, the British physician and consumer activist, has condemned this practice in no uncertain terms: "The pallid yolks of commercial eggs can legally be colored with any yellow dye, however dangerous, if, being fed to the hen, it is excreted into the yolk. This deluding the public by providing battery eggs with yolks to the golden yellow of 'farm' eggs is a dangerous swindle and should be banned."[16]

To induce increased production, egg farmers manipulate lighting systems so that the laying hens will believe three days have passed instead of two. The egg yield is further increased by "force-molting" the birds—speeding up the process that hens undergo after a certain period of egg-laying. This is accomplished by dehydrating the birds, keeping them in complete darkness for a week, and doctoring their feed with molt-producing hormones. Hens subjected to this treatment often lay faulty eggs that are cracked or misshapen, but the processor can still sell them to restaurants and industrial bakeries. By the time her laying days are over, the old hen is literally worn out, her flesh emaciated and her bones brittle. She is not even fit for the stew pot, and often only pet food manufacturers will have her.

Before reaching your supermarket shelves, battery eggs are washed with water, detergents, and disinfectants to cleanse them of excrement and other debris. To extend their shelf life, the eggs are dipped into or sprayed with an oil treatment that is virtually undetectable to the consumer. Since the shell of an egg has about 7,000 pores, a quantity of these substances are absorbed. The F.D.A. does not consider this contamination and they never test eggs for such residues.

The "problem" eggs that you never see (but may nevertheless eat) are the imperfect and cracked ones, which are also known as "floor" eggs because they have slipped from the nests and are frequently broken. These may be culled from the rest and shipped to manufacturing plants in cold storage. Egg-cracking machines automatically separate the yolks from the whites, and pour them into containers that are flash-frozen and shipped to bakeries and food processors. Fortunately, the person shopping at a market can examine eggs for hairline cracks, and thus avoid the danger of being infected with salmonella. At the same time, he or she may buy a convenience food made from contaminated eggs—and there is no way that anyone can determine its safety before the food is eaten. Usually, the high heat required in food processing will destroy salmonella bacteria, but not always. Products containing industrial-grade frozen egg whites, including meringues, cream pies, angel food cakes, and imitation mayonnaise, are particularly questionable, because the heat necessary to kill salmonella would also destroy their cooking properties.

In 1966, the U.S.D.A. hoped to contain this problem by requiring that all products distributed nationally be made with pasteurized egg whites. The ruling did not cover eggs and egg products made locally and not shipped across state lines, however, and consequently only well-known national brands are constantly safe to eat. In April, 1967, 1,800 people in New York City contracted salmonella as a result of eating an imitation ice cream dessert made with unpasteurized frozen eggs.[17]

Although they cost more, the most nutritious eggs you can find are at your health food store. They are eggs produced without chemicals or drugs from chickens that roam free. Most have made the acquaintance of a rooster, and their eggs are living organisms, with the vital enzymes that are present only in living things.

A Real Turkey

All the information presented on these pages regarding chickens applies equally to turkeys. These magnificent birds, which are as much a symbol of America as the bald eagle, have been turned into pre-packaged commodities that are now available the year around, thanks to deep-freezing and chemicals. The greatest demand still comes during the Thanksgiving and holiday season, however, at which times fresh birds seem to be abundantly available. Actually, they could be frozen birds slaughtered months before to level out production schedules, and released during the period of peak demand. Technologists have made it possible to treat a turkey before freezing so that its thawed skin looks normally pink, instead of a tell-tale reddish-brown characteristic of a frozen bird.

Regardless of the claims of turkey processors, freezing has never been known to improve any food, and frozen turkeys tend to be dry, stringy, and flavorless. In recent years, manufacturers have come up with a highly promotable and specious merchandising gimmick designed to overcome these deficiencies. The self-basting turkey is usually advertised as possessing a "buttery" flavor, which (unlike the name suggests) could be supplied by an artificial flavor and a solution of coconut oil and water. As highly saturated as lard and twice as cheap, coconut oil turns lean turkey meat into a less healthy protein source, especially for those on low-cholesterol diets.

Another popular form of processed turkey is the frozen turkey roast, which is made of pieces of turkey rolled and tied, and ready to cook as is. Most of the meat is taken from whole birds, but some of it is salvaged from damaged or diseased ones. If that isn't discouraging enough to you, take a look at the additional ingredients: sodium tripolyphosphate (a water-retainer that holds in natural juices), brown sugar, sodium hexametaphosphate (a water softener used in Calgon), sodium erythorbate (an antioxidant that preserves freshness), and salt.

Learn about these products and seek out independent butchers, who are more likely to carry fresh turkeys—at least during the holiday season. When more consumers begin to demand fresh, healthy, flavorsome birds, supermarkets and poultry raisers will be forced to supply them.

Pork—Adding Injury to Insult

Hog-raising has not escaped the weight-forcing process, either. Crowded into filthy sties, pigs are given tranquilizers in their drinking water to prevent them from biting each others' tails, a neurotic reaction to stress that rarely occurs on traditional farms. Like many a modern city dweller, about half of all American pigs develop stomach ulcers due to stress and an improper diet. In many states hogs are fed processed garbage that is cooked first over high heat, which destroys a number of essential amino acids and vitamins.[18]

The garbage diet plus poor sanitary conditions in which most hogs are kept add to the number of diseases that hogs contract. One result is a high mortality rate among piglets, a profit-destroying problem that can only be solved by taking them away from the sow shortly after birth. The piglets are put into temperature-controlled pens where they are fed a synthetic diet rich in vitamins and minerals. Given the care and nutrition that nature is no longer able to supply, most of the weaklings pull through. Meanwhile, the sow is injected with a chemical that induces heat and makes it possible for her to mate almost immediately. The man-made process turns her into a virtual breeding machine that produces three yearly litters instead of the customary two.

In some modern piggeries, the animals are crowded together so tightly they can barely move. Moving around uses up energy that burns body fat, and hog raisers want their animals to be as fat as possible. Hormones, antibiotics, and copper sulfate, a cumulative poison, are added to their feed to make them grow fatter faster; tranquilizers calm them into immobility; and antibiotics prevent the spread of infectious diseases. All these treatments are simply an indication that modern methods of hog raising are hopelessly out of kilter with the laws of nature. By ignoring the basic needs of their pigs and depending upon chemicals to pull them through and get the animals to market, the hog farmers are producing meat that is a mere shadow of what it once was.

Livestock scientists and meat packers refer to the present poor quality of pork as P.S.E., which stands for pale, soft, exudative (water-logged) flesh. It has been estimated that one out of every five pork cuts available at the supermarket fits into this category—the result of growth stimulants and a poor diet that often includes empty peanut hulls and ground-up newspapers.

After a short life spent in a crowded environment, eating processed garbage, the pig is subjected to a final insult before slaughter. Sodium pentobarbital, a powerful anesthetic, is injected into its bloodstream to relax its muscles, so that the flesh will be redder and more tender. Dr. Jonathan Forman, former editor of *Ohio State Medical Journal*, has written a fitting epitaph: "We know . . . that the typical pig ready for market is a sick animal—the victim of obesity—who would die long before its time, if we did not rush it to market for the city people to eat."[19]

Because there is presently no requirement for inspection of pork, and no grading standard, the consumer buys at his or her own risk. Trichinosis, a round worm larva, is contained in some pork. If the meat is not cooked to an internal temperature of at least 170°, the trichinosis will not be destroyed, and the parasitic larvae may attach themselves to your intestines and reproduce. If this happens, new larvae will migrate to muscle tissue and encyst, causing a grave health problem that is difficult to cure.

The Sham Ham

The ham our grandparents ate was cured naturally by salts, a little sugar, spices, and herbs. Applied to the surface of the meat with a little water, these organic ingredients combined with the ham's natural juices to form a brine solution in which it was soaked for two months. After being cured, the ham was then smoked over burning hickory logs to dry it out and flavor it. The time-consuming method produced delicious hams that bear about as much relationship to today's supermarket product as china cups do to plastic ones.

The decline in flavor and quality began in the 1930s, when processors invented a shortcut for curing that enabled them to add ham to the growing list of adulterated, mass-produced products. Instead of soaking the meat, a brine solution was injected into the main artery of the ham and pressure-pumped throughout the tissue. Known as "pumping," this instant curing technique also enabled processors to cheat by overpumping hams with water to increase their scale weight. Quite rightly, the U.S.D.A. eventually banned overpumping, which had become a major problem by the 1940s. At the same time— with the F.D.A.'s approval—the U.S.D.A. inexplicably granted permission to add chemical phosphates to the injected ham. Phosphates increase the amount of water absorbed into the meat without making it look wet or water-logged.

Nevertheless, the packing industry was not satisfied, and they petitioned the U.S.D.A. for permission to pump even more water into hams. The U.S.D.A. established and approved legal tolerances for the additional water, claiming that "consumer demand for juicier smoked meats is increasing." No public hearings were held and the decision was withheld from the press, making it obvious that the government agency was blatantly abusing its position of public trust.

In 1960, the fledgling consumer movement learned of this consumer sell-out and President Kennedy's Secretary of Agriculture, Orville I. Freeman, was pressured into calling open public hearings on the matter. One women's organization characterized the U.S.D.A.'s permissiveness as "governmental approval of cheating," while others pointed out that they and their families were paying high prices for water while being shortchanged on nutrition.[20] On October 31, 1961, Secretary Freeman rescinded the U.S.D.A.'s order, but Armour & Company—the industry giant—took its case to court and won. A compromise was reached and packers were permitted to add up to 10 percent water to hams as long as the words "water added" were included on the label.

That ruling is still in effect today, and it applies only to plants that do interstate business; local plants are outside federal jurisdiction and their hams often contain up to 30 percent water. None of these figures, much less the label, include the 2 percent or more of chemical additives injected into ham, including phosphates, artificial smoke flavoring, sodium nitrites, and sodium nitrates.

Frankfurters and Cold Cuts—
Everything but the Squeak

Everyone who eats frankfurters and cold cuts should

take a tour through a local packing house. The meat that goes into hot dogs and luncheon meats can sometimes be the most repugnant parts of cattle, pigs, and chicken. Among the ingredients are pork jowls, salivary glands, pork stomachs, lymph nodes, pork spleens, pork lips, and pork cheeks—everything but the squeak. Poultry meat could be included in these formerly beef-and-pork products, and the U.S.D.A. permits additions of up to 15 percent without requiring notice on the label. Thus, the poultry industry has a convenient market for "spent pullets"—worn-out laying hens that are worthless to anyone but meat processors and pet food manufacturers.

Frankfurters, knockwurst, and bologna are permitted to contain up to 30 percent fat, 10 percent added water, 3.5 percent meat extenders (cornstarch, soy flour, sugar, or powdered milk), and a list of additives so long that it is impossible to include them here. Unless labeled "all beef," these products may contain pork, mutton, lamb, goat, turkey, and chicken. Sausage is even more unwholesome, for it contains up to 50 percent highly saturated pork fat. Moreover, conditions are so unsanitary in some sausage plants that the F.D.A. has found rodent hairs, mouse excreta, and insect bodies in random samplings of even major brands.

The little bits of real meat that go into hot dogs and bologna are those that happen to be attached to fat trimmed from chops and hams. All this meat offal is pulverized into a thick paste with a chemical emulsifier, and water, fillers, and other additives are added to the huge cooking vats. At this stage the mixture looks like a gray bread dough and it has the consistency of mud. The pasty mush is next forced into miles of casing coated with a brilliant red coal-tar dye, tied into links, and cooked. The action of heat combined with additives like sodium erythorbate and sodium acid pyrophosphate force the color into the mixture, and inferior ingredients are transformed into a product that looks and tastes good.

The Failure of Meat Inspection

There are three major categories of meat-judging in this country, all of which fall short of their original intentions. The first and most important of these is the federal inspection by the U.S.D.A. of all meat sold across state lines. State inspectors handle this task for locally produced and sold meats and meat products. The third category is the government grading system, which is a voluntary system agreed to by the slaughterhouse.

With the exception of government controls on banks and railroads, meat inspection is the United States' oldest regulatory system. The first laws were passed in 1895 requiring that animals be inspected for diseases such as tuberculosis before slaughter. The first law requiring mandatory inspection of meat before, during, and after it is slaughtered was passed in 1906. Then, as with the Wholesome Meat Act passed 61 years later, the law came about only after groups of concerned citizens joined together to reveal the facts and fight the powerful meat-packing interests.

Led by crusader Upton Sinclair, who described the filthy conditions that existed in Chicago's stockyards (among the nation's largest) in his powerful book, *The Jungle*, this nameless, vestigal consumers' group was one of the first to mobilize the public to act in its own favor. The popular muck-raking press of the day picked up the Chicago story, and readers were shocked to learn that diseased animals were not culled from healthy ones, and that horsemeat and rats were added to sausage hoppers as a matter of course. The most horrifying story of all concerned a young worker who allegedly slipped and fell into a vat of boiling meat by-products slated to be made into cold cuts. Rather than stop the machinery to remove his remains, the management decided too much time would be lost, so he was included in the next day's bologna shipment. President Theodore Roosevelt hastily signed into law the Meat Inspection Act to "insure wholesomeness from the hoof to the can."

For the next 60 years, the public believed that the problem had been solved while they unknowingly continued to buy meat of questionable quality. One basic problem, it soon became apparent, was that meat inspectors were not adequately enforcing the law, and that some were being paid off by packing plants. Meat inspectors were supposed to check for sanitary conditions in the slaughterhouse, proper labeling and packaging. It was cheaper for packers to pay off underpaid inspectors than to adhere to expensive, time-consuming regulations. More important, there was a loophole in the law, which covered only meat sold in more than one state. In most states, therefore, processors not engaged in interstate commerce were free to do as they wished.

In 1963, the U.S.D.A. surveyed many of these state meat packing plants and uncovered conditions so deplorable that its report was suppressed until 1967. In Delaware, for instance, the report revealed: "Rodents and insects, in fact any vermin, had free access to stored meats and meat product ingredients. Hand washing lavatories were absent or inadequate. Dirty meats contaminated by animal hair, the contents of the animal's digestive tract, sawdust, flies, rodents and filthy hands, and the tools and clothing of food handlers were finely ground and mixed with seasonings and preservatives. These mixtures are distributed as ground meat products, frankfurters, sausages, and bologna. Due to the comminuting [pulverizing] process and seasoning of these products, most of the adulteration could not be detected by the consumer."[21]

Spot checks around the country conducted several years later revealed that sanitary conditions had declined in federally-inspected plants as well. A U.S.D.A. inspector found that workers in a North Carolina packing house would "spit on the floor, then drop sausage meat in the same spot, which was then picked up and shoved into the stuffer." Another U.S.D.A. employee visited a plant in Norfolk, Virginia, where he found "abscessed beef and livers, abscessed pork liver, parasitic livers mixed with the edible product." Concerned employees of the Department of Agriculture leaked this secret information to Ralph Nader, other consumerists and citizens' organizations, and influential newspaper reporters, all of whom began to lobby for the passage of new legislation to further protect the public.

Hearings were held in the House of Representatives, and the meat industry fought back with a battery of highly paid lawyers, a massive public relations campaign, and a powerful Washington lobby that had garnered the support of several Senators and Congresspersons. The evidence was overwhelming, however, and nothing the industry could do was powerful enough to cover the stench emanating from the deplorable conditions they had sought to hide. Newspapers daily reported shocking new revalations as they surfaced. Ranchers told Congressional investigators that state plants were their only outlet for getting rid of what was known in the trade as "4-D meat"—meaning dead, dying, diseased, and disabled. It soon became clear that limiting business to the states of manufacture was merely a ploy by meat packers and processors to avoid federal inspection practices (when they were actually conducted). Many industry giants, whose names were household words and therefore thought to be federally inspected companies, were in reality doing business state-by-state in order to save money by using "4-D" meat.

Government support for the meat industry's position began to collapse as their many abuses came to public attention. In a 1967 magazine article, Ralph Nader compared current conditions to those that had led to the passage of the Meat Inspection Act of 1906. "As far as impact on human health is concerned, the likelihood is that the current situation is worse," Nader wrote. "The foul spectacle of packing houses in the earlier period has given way to more tolerable working conditions, but the callous misuse of new technology and processes has enabled today's meat handlers to . . . cover up meat from tubercular cows, lump-jawed steers, and scabby pigs. Chemistry and quick-freezing techniques provide the cosmetics of camouflaging the products and deceiving the eyes, nostrils, and taste buds of the consumer. It takes specialists to detect the deception. What is more, these chemicals themselves introduce new and complicated hazards unheard of sixty years ago."[22]

Finally, on December 15, 1967, President Lyndon B. Johnson signed the Wholesome Meat Act into law. The victory became a touchstone of the consumer movement, and valuable lessons were learned about fighting strategies that gave the movement new strength. By pitting the government against the meat and farm industries on the issue, they had employed the age-old but effective strategy of divide and conquer. In light of the well-publicized disclosures, the government had no choice but to defend the interests of the people who had elected its members. In holding back information until the moment of maximum impact, the consumerists caught meat industry representatives off guard and completely upset their line of defense. Once and for all, the victory proved that if we the consumers were given all the facts, nothing could prevent us from generating the political pressure needed for reform.

The new law made it mandatory for all states to legislate inspection and quality standards by 1970, consistent with federal specification. Yet, even if inspection methods are improved 100 percent, the law falls far short of being a guarantee of safety and wholesomeness. For one thing, all the antibiotics, hormone residues, and additives are still there, and inspection practices continue to be lax, primarily because there are not enough meat inspectors to do the job.

Under the best of conditions, meat inspection is a rough and dirty job, with the inspector being constantly pressured by plant managers and owners. Most inspectors are veterinarians by profession, although they are not required by law to be so, and they are expected to handle animals being brought to slaugherhouses, as well as their dead carcasses. They must also smell, feel, and closely examine all organs and tissues, including the liver, heart, kidneys, spleen, lungs, air sacs, sex organs, and fecal matter.

Usually, there is not enough time to do this properly, since they may have only seconds or minutes for each carcass or chicken. Additionally, some diseased animals are difficult to diagnose since the tumor or afflicted tissue may be concealed in an inaccessible spot. Chickens, for example, often are contaminted with leukosis, avian cancer, and they are supposed to be taken off the market because they are a possible threat to human health. It is almost impossible, however, to spot the microscopic tumors produced, and consequently a great number of diseased chickens slip by the inspectors. Another chicken disease is airsacculitis, somewhat like human pneumonia, in which pus-laden mucus collects in the chickens' lungs. The U.S.D.A. permits the sale of such chickens provided the lungs are cleaned out with air suction guns, a practice that often breaks the sacs, releasing the pus throughout the chicken. The result is evident in supermarket storage coolers, where grayish and yellowish skin and tumorous lumps bear witness to the illnesses of these unfit birds.[23]

With such a difficult job to perform, working in noisy plants where one must shout to be heard a few feet away, it is not hard to imagine why a meat inspector might not be able to do his work effectively all of the time. Factory owners know that the inspector has real power to not only condemn diseased meat, but also to suspend operations; this has led to a few unscrupulous owners trying to bribe, threaten with violence, or influence the inspectors' superiors with investigation of their private lives, with the threat of blackmail implied. The most recent incident of industry payoffs to meat inspectors occurred in New York City in April, 1976. According to the District Attorney, the investigation began when agents of the F.B.I. discovered that there existed a routine system of bribes in meat packing plants that had existed for years. The payments were believed to be $10 to $100 per week, including gifts and a regular supply of meat. Fifty-two meat inspectors were subpoenaed by a grand jury. Because of this chicanery it is anyone's guess how many tainted pounds of meat and poultry were approved and found their way to the consumer.[24]

If meat reaches the supermarket uncontaminated, however, it is not guaranteed to stay that way. A team of

reporters from the *Philadelphia Bulletin* visited a number
of markets in 1973, gathering information on cleanliness.
One meat department worker confessed to a reporter that
he was forced by his supervisor to doctor 100 spoiled
chickens that he had wanted to throw out. "They smelled
just like rotten eggs," the clerk said, "and the air was thick
with odor. [The boss] told me, 'When I get through with
these you won't know they're bad.' He filled up the sink
with cold water, dumped three boxes in, and covered
them with baking soda. He mixed the baking soda around,
pulled the plug out, drained the water, and bagged the
chickens. We sold them within a few days."

Another worker revealed his store's method of getting
rid of old hams. "The hams were slick with bacteria and
green and white mold," he told a reporter. "Many of the
hams had dark spots on them where the mold had eaten
through to the tissues. We washed them off with scrub
brushes in tubs of water and got off the mold, but you
could still see the damaged spots, and they smelled awful.
I didn't see how anyone would buy them. Then the boss
poured a couple of bottles of liquid smoke flavoring over
them, and you really couldn't tell they were spoiled. Those
we couldn't sell we later chopped up for making ham
salad."[24]

It is difficult to let your eyes and nose be the guide to
buying fresh meat these days, especially when it is packed
in plastic containers where unsavory sections are hidden
under the attractive part. Once you get the package home,
you find that you have purchased unwanted bone, gristle,
and fat—or worse. Until the battle is won and meat is once
again uncontaminated with chemicals and filth, I recommend you stay away from it in favor of fish, cheese,
soybeans, and other sources of vegetable protein. If you
can't live without meat (you can if you try), your most

trustworthy source is the butcher who will let you see the meat before it is wrapped and sealed. Remember that fresh beef is cherry-red in color, its texture is velvety, not coarse, and the fat portions should be pale—not deep—yellow. Fresh poultry will feel firm to the touch; it will have no odor (not even a chemically-induced "fresh" one), and its skin should be resilient enough so that your finger will not make an indentation in it.

Notes

1. U.S. Department of Agriculture, Economic Research Service, 1975.
2. Bundy, Clarence E., and Duggins, Ronald, V., *Livestock and Poultry Production*, Prentice-Hall, Englewood Cliffs, N.J., 1968.
3. Zwerdling, Daniel, "Beefed-up: Drugs in the Meat Industry," *Ramparts*, 1972.
4. Boyland, E., *Correlations of Experimental Carcinogenesis*, Experimental Tumor Research (E. Homberger and Karger, editors) Basel, 1969, pp. 222-34.
5. Zwerdling, *op. cit.*
6. Nelson, Bryce, "Welfare Department Eliminate Blacklist of Science Advisors," *Los Angeles Times*, January 3, 1970.
7. "Dangerous Foods and Drugs," *U.S. News and World Report*, February 23, 1976.
8. *The New York Times*, April 7, 1976, p. 48.
9. *The U.S. Fact Book—The American Almanac*, Grosset & Dunlap, New York, 1976.
10. *Ibid.*
11. Cottoam, "Pesticide Plan Dangers Cited," *Albuquerque Journal*, July 11, 1969.
12. "Antibiotics on the Farm—A Major Threat to Human Health," *The New Scientist*, October 5, 1967.
13. The National Communicable Disease Center, *Annual Report*, 1975.
14. Meat & Poultry Inspection Program, *Regulation of Food Additives and Medicated Animal Feeds*, p. 421.

15. Gass, George, "Carcinogenic Dose-Response Cure to
    D.E.S.," *Journal of National Cancer Institute*, Vol. 33, No.
    6, December, 1964.

16. U.S.D.A. Yearbook of Agriculture, *Consumers All*, 1965.

17. Bernarde, Melvin A., *The Chemicals We Eat*, McGraw-Hill,
    New York, 1975.

18. *Successful Farming*, December, 1962.

19. Longgood, William, *The Poisons in your Food*, Simon &
    Schuster, New York, 1960, p. 159.

20. *U.S.D.A. To Reopen Ham Study*, USDA release, March 17,
    1961.

21. Nader, Ralph, "Watch That Hamburger," *The New
    Republic*, August 19, 1967.

22. Nader, Ralph, "The New Meat Law," *The New Republic*,
    July 15, 1967.

23. "Replacement Pullets—Their Feeding, Rearing," *Feed
    Age*, July, 1965.

24. *The Philadelphia Bulletin*, June 10, 1973.

Pretty Poisons—
The Ugly Face of Beauty

After polluting our bodies with synthetic foods, chemical residues, harmful additives, and noxious gases suspended in the air, is it any wonder that our exterior physical appearance begins to suffer? A look in the mirror reflects hair that has become thin and lusterless, a complexion gone dull and ashy, or pimples and rashes on various parts of the skin. Our bodies are paying the price for constantly ingesting foreign substances and we are beginning to look old before our time.

Rather than correct the symptoms of deteriorating health with proper diets and simple cleansing and maintenance procedures, most people buy cosmetics that contain more dangerous chemicals, and try to mask the damage. In an attempt to restore the appearance of healthy good looks, people rub, pour, sprinkle, and spray themselves with cosmetics, unaware they could be making matters worse.

Cosmetic manufacturers—who are one of the largest advertisers on television and in magazines—relentlessly pressure us to conform to fabricated standards of beauty by using products that may only be functionally worthless. People recoil in disgust at primitive peoples who achieve their beauty by bathing their faces with urine and pomading their hair with cow's dung, completely unaware that they are doing even worse to themselves.

Playing on the fear of social rejection and appealing to vanity, cosmetic advertisers urge one to clog the perspiration ducts with metal salts so as "not to offend" with natural odors; to apply caustic acids to "revitalize" the hair; to paint one's face with multiple coal-tar dyes for the "dewy look of youth"; and to scrub the enamel from one's teeth because "whiter teeth means more sex appeal."

The promise of a shortcut to attractiveness is a highly effective sales pitch, and American consumers respond by spending in excess of $14 billion yearly[1] to paint, powder, and "pretty" themselves. The markup of such "necessities" as face creams, foundation liquids, and lipsticks is usually 500 to 1,000 percent above the cost of manufacture,[2] which is much less than the amount spent on advertising and packaging the products. The cosmetic companies are in business to promote the illusion of health—and the real cost is often paid by the consumer, not at the counter but in his or her body.

The Skin Game

The skin is a many layered living cloak of approximately 20 square feet, and it contains a complex network of nerves, blood vessels, glands, and cells, including the hair and nails. Tough and resilient, this largest organ of the body protects the internal organs, regulates temperature, helps rid the body of wastes in the form of perspiration, and thrives on a minimum of care.

Toxins in the system show up first in the skin, the appearance of which reveals a great deal about what has gone into the body.

The skin is nourished from the inside out, and, despite the claims of cosmetics' manufacturers, no creams can be absorbed to any great extent or effectiveness. To maintain the glow of health, you must carefully and vigilantly avoid the poisons described in this book, and maintain a natural, balanced diet that will provide all the nutrition your skin needs.

Any foreign substance applied to the skin may interfere with its natural processes and could cause allergic reactions. Contact dermatitis, the most common, can range from tiny pink spots to angry bright red oozing patches covering large areas of the body. Acute reactions, although admittedly rare, can occur within seconds of application and result in shortness of breath, heart failure, and even death. Cosmetic ingredients, such as hormones and antibiotics or the highly touted hexachlorophene, have been proven harmful to everyone. You not only run the risk of suffering from an allergic reaction yourself, but the cosmetics you use may induce an unexpected physical reaction in a friend or a lover, who may be allergic to what *you* are wearing.

Cosmetic companies only now admit that simple cleanliness and the use of an emollient cream such as soy oil as needed will do more for your appearance than many of their adulterated products. The following list offers safe alternatives to name-brand products and they are all readily available. After years of delay, the F.D.A. finally passed a ruling in 1975 that required manufacturers to list the ingredients in their cosmetics and now, by reading the label, you will know which products to avoid.

Soaps, Detergents, and Cleansing Creams

Proper cleansing is essential for a clear, radiant complexion. Dead cells, rancid oil, perspiration wastes and bacteria must be removed daily or blackheads and pimples will form in clogged pores. By the end of a work day this film of natural detritis will also contain dirt, airborne pollutants, and—if you wear it—stale makeup.

Soap is the oldest and easiest method of removing surface grime and also the safest. A combination of fats, alkali salts, and water, soap has the double advantage of cleansing thoroughly and rinsing itself from the skin. Detergents lather better, especially in hard water, but they may irritate dry skin and therefore should be used only by those with oily skin or acne conditions, and even then with considerable caution.

Unfortunately, many manufacturers have chosen to ruin a perfectly good bar of soap by coloring it with potentially harmful dyes and chemical perfumes. Choose a white, unscented bar such as those available in health food stores, or stay with *Ivory*, which is the closest you can come to an unadulterated product in your supermarket soap aisle. Dry skins may require a super-fatted bar such as *Basis*, which contains additional oils to make it milder.

Cold creams and cleansing creams should be avoided entirely, as they leave an oily film no matter how well they are tissued off, and the skin is only partially cleaned. The basic ingredients of mineral oil, wax, borax, and soap or detergents react negatively with sebum (natural oil secretions), and can cause allergic reactions.[3]

Deodorant Soaps and Hexachlorophene

Until the F.D.A. banned its widespread use in 1972, hexachlorophene was added to some 400 cosmetics ranging from soaps to vaginal deodorants. Touted as a new "miracle" ingredient that "stops b.o. (body odor) before it starts," the chemical destroyed bacteria that converts normally odorless perspiration to its characteristic pungent state. At the same time, the antiseptic remained on the skin where it slowly worked its way into the bloodstream.

After years of allowing hexachlorophene to be indiscriminately applied to every part of the body, the F.D.A. belatedly discovered that the substance caused brain damage in laboratory rats and monkeys, which indicated that it may have the same deadly effect on other mammals, most notably humans. An immediate concern was its exposure to babies, who were bathed regularly in the germ-killer by hospital nurses, a practice then perpetuated by their unwitting mothers. No infant brain damage has yet been reported, but it is well within the realm of possibility that affected infants have been misdiagnosed, and their mental retardation laid to other causes.

Manufacturers hastily substituted other inadequately tested chemicals to kill bacteria on the skin, primarily bithionol, which is closely related to hexachlorophene, and has been known to cause skin rashes and swelling. Be very sure you carefully scrutinize the ingredients listed on a cleansing product and avoid using any which contain these harmful additives.

Deodorants and Antiperspirants

Perspiration is secreted from two types of skin glands, the eccrine and apocrine sweat glands. The eccrine glands are found over most of the body and they produce a mild secretion. It is the secretion of the apocrine glands, which exists in the armpits and genital region, that generates a fetid odor when it interacts with bacteria nestled in hair follicles.

Deodorants are designed to eliminate the scent of apocrine perspiration by eliminating the bacteria that cause it. The effective ingredients include bithionol and a mild antiseptic such as alcohol. There are also many deodorants that depend upon odor substitution and mask a disagreeable smell with a strong perfume. Antiperspirants eliminate perspiration odor by stopping its flow with aluminum salts that swell the pores shut, effectively preventing sweat from reaching the skin's surface. Both are harmful and are the leading causes of cosmetic injuries in the United States.[4]

Danger signals range from itching rashes, blisters, and lesions to tiny, hard, long-lasting lumps in the armpits, which are known as granulomas. To make matters worse, the cosmetic industry packages deodorants and antiperspirants in aerosol spray cans that propel these ingredients into your lungs where they can remain indefinitely.

Recently, two young women turned up at the U.S. Army's Fitzsimons General Hospital in Denver, complaining of fatigue and chest pains. Doctors discovered that the women, who were roommates, both used aerosol deodorants which had badly inflamed their lungs. A rare disease called sarcoidosis resulted and the women were promptly hospitalized. Doctors at Fitzsimons discovered that similar cases had been reported and they undertook a laboratory experiment with several popular deodorant brands. Guinea pigs exposed to the various commercial sprays all developed lung lesions.[5]

If soap and water aren't sufficient to keep you comfortable and fresh smelling then use a natural substance to deodorize your apocrine glands. Try a light dusting of baking soda, or a preparation that contains fuller's earth, a clay product obtainable at your local health food store; neither is toxic.

Vaginal Deodorants

Perhaps the most useless products ever concocted by the cosmetic business are "feminine hygiene sprays," the euphemism for women's genital deodorants. With the arrival of the so-called sexual revolution in the 1960s, the beauty trade decided it was time for a genital cosmetic and before long women had 30 brands to choose from "to prevent intimate odor."

Most women produce a normal, mild secretion of the external genitals, which may develop an inoffensive odor when the body has been confined in pantyhose or a panty girdle all day. Normal bacteria act on the secretion and perspiration that has been unable to evaporate. Spraying chemicals and propellant gases on the delicate mucous membranes of the vagina is no solution, and it can in fact be dangerous. They have been known to cause bladder infection, vulval itching or burning, boils, and blood in the urine in addition to other minor irritation.

The F.D.A. reports that complaints are running at ten times the rate of what is considered normal for cosmetics. Yet, despite mounting evidence and pressure from the Consumers Union and other important citizens' groups, the F.D.A. has so far refused to reclassify the sprays as the drugs they are or even stipulate cautionary labeling. As such, manufacturers would be required to submit these dangerous products to controlled tests in order to prove their effectiveness and verify their safety.

The highly respected drug-evaluating publication, *The Medical Letter*, offers the following common-sense advice:

"It is unlikely that commercial feminine hygiene sprays are as effective as soap and water in promoting a hygienic and odor-free external genital surface."

Moisturizing Creams and Lotions

Whether they are called night creams, eye creams,

moisturizers, hand creams, or body lotions, all these products are intended to perform essentially the same function and they contain basically the same ingredients. They are emollients, which means that they smooth and soften the skin and form a protective layer that holds in moisture. Such creams usually contain mineral oil, lanolin, beeswax, and natural oils such as olive, peanut, or sesame. These ingredients are fine in and of themselves, but manufacturers proceed to add dangerous chemicals in order to prolong the shelf life and give color and attractive scents to their products.

The worst of these is a family of preservatives known as paraben esters. Because most creams are not packaged under controlled sanitary conditions, parabens are mixed in to kill bacteria that may be present. The *Journal of the American Medical Association*[6] reported an alarming number of cases of violent reactions to this fungistatic agent, including chronic eczema, crusty lesions, red and painful eyes, and swelling of the eyes and face.

Perfumes present in face creams can also cause contact dermatitis. In fact, fragrances in any form should never be applied to the face, which is the most sensitive area of the body.

Oily skins require no creams or moisturizers whatsoever, but normal and dry skin will benefit from a light application after washing. Lanolin closely resembles natural human sebum (oil) and possesses the added advantage of not interfering with the skin's normal activities, such as sweating and eliminating the natural waste products of the skin's constantly replenishing cells. Ask your druggist for a tube of ordinary toilet lanolin—which is probably hidden behind a stack of expensive additive-ridden pseudo-creams—and you will be ahead of the skin game. Alternatively, you can moisturize with Vaseline, which is nearly as good.

"Youth" Creams and Hormones

Ever aware of the human desire to stay young forever, cosmetic companies promise us that wrinkles will vanish and chins will no longer sag if we will only buy their turtle oil, royal jelly, and/or unborn lamb extract. None of these substances has had the least effect on aging skin, however, so manufacturers have suggested a newer, more dangerous panacea—hormones.

Applying creams containing hormones to our skins can cause severe physiological changes in living tissues. At a time when medical doctors are having second thoughts about prescribing estrogen (female) hormone treatments for women with post-menopausal discomfort because of the synthetic drug's link with cancer of the cervix and uterus, the F.D.A. nonetheless continues to permit its use in cosmetics. All hormone creams should be carefully avoided until more is learned about the immediate and residual effects they may have on our skin and our systems as such creams are absorbed into the body.

Foundation Makeup

No makeup is beneficial to the skin, including the hypoallergenic kind. A mixture of oils, fatty acids, clay pigments, and sensitizing preservatives, foundation makeup clogs the pores as it covers the skin and attracts pollutants from the air. Removing it effectively may cause you to rub too hard and damage the collagen fibers below the skin's surface and thus hasten the aging process. Makeup also contains foreign material that cannot help the skin in any way and is a potential source of harm.

Medicated makeup is even worse, and its prolonged use can cause significant problems with the body's immunological defense system. Used primarily by those with acne for the purpose of covering up pimples while curing them, medicated foundations do offer temporary relief, but the ultimate damage they have been known to cause is a high price to pay for a provisional cure.

By applying antibiotics and antibacterials to the skin, you are building up a resistance to them. Then, in the case of an emergency, such as a wound or massive bacterial invasion of the body, antibiotic injections may not be as effective. Acne can only be helped by a proper diet low in fats and sugars, regular and proper cleansing, and application of plain rubbing alcohol.

Powder and Talcum

Face powder is called "the finishing touch" and is really used to blot out the shine from oils in foundation creams. Its principle ingredient is talc, a soft mineral known as hydrous magnesium silicate, but it also includes clay pigments, zinc oxide, and perfumes. Talcum powder is composed of essentially the same ingredients and therefore presents the same dangers.

Similar in composition to asbestos and mined from adjacent veins, talc has been found to contain dangerous levels of this mineral.[7] Since it is impossible to use a powdered substance without inhaling it, most regular users could be harboring large quantities of these indestructible minerals in their lungs. An asbestos fiber or talc particle you breathe at 16 will probably stay in your lungs until you die. Less is known about the long-term effects of pure talc, but asbestos poisoning can result in cancer of the lungs or mucous membranes, and slow suffocation from the hard scar tissue formed in the lungs as they fight helplessly to slough it off.

Cornstarch, an organic substance that can be absorbed by the lungs if inhaled, is the *only* safe alternative.

Lipstick and Rouge

Essentially a tube of oil mixed with wax and a coal-tar dye, lipstick is applied where it can do the most harm to the body. Unprotected by the horny layer of skin that covers the rest of the body, the lips are highly susceptible to allergic reactions caused by the indelible dyes used in most lipsticks. Cheilitis—contact dermatitis from lipstick—is the most common complaint, the symptoms of which are dryness, chapped and cracked lips, and swollen gums.

The culprits are the coal-tar dyes, which are also ingested into the body. Almost all women who wear lipstick must re-apply it several times a day because they inadvertently lick it off when talking and eating. These minute amounts of lipstick have been known to cause gastrointestinal upsets like gastritis and colitis. The F.D.A. blandly ignores this fact. While it prohibits coal-tar dyes in eye makeup because they *might* enter the body, they allow them in lipstick where they are sure to enter the body.

The cosmetics industry attempts to justify the continued use of these dyes by claiming that there has never been a proven case of anyone being poisoned by lipstick. But how many doctors examining a patient with blood or liver damage caused by coal-tar products would suspect lipstick? It is a clear and established fact that coal-tar colors are dangerous to health, and that their effect is cumulative. It is best to refrain from using any lipsticks since there is no law that requires the listing of ingredients.

Rouge and blush are variations of lipstick and since they are placed on the cheeks, rather than the mouth, to impart a healthy glow they could be considered not as dangerous. A natural glow is the most beautiful glow, but if you feel you must use these products, at least buy those you can determine are colored with vegetable dyes. Large department stores and health food stores stock them.

Mascara and Eye Makeup

Mascara is a soap or cosmetic wax with lampblack pigment that is used to darken eyebrows and eyelashes, and hence make the eyes appear larger and more beautiful. Eye makeup is one of the leading causes of cosmetic injuries[4] in the United States and mascara is the worst offender. Fortunately, the eye can tolerate little irritation and users of mascara quickly abandon a product that causes smarting, tears, and bloodshot eyes. Mascaras containing "lash extenders," intended to make the eyelashes appear longer and fuller, can have more serious consequences. These products contain nylon fiber that adheres to the lashes and can easily fall into the eye and scratch the cornea. They are especially unsafe for wearers of contact lenses.

Eyeliner pencils are sometimes used to outline the upper and lower borders of the eyelids, a practice advised against by the American Medical Association. Their files are filled with cases of women who have lacerated the tender mucous membranes around the eye, colored them permanently, and impaired their vision.

The F.D.A. reports similar cases resulting from the use of false eyelashes, and at least one instance of permanent blinding from the adhesive used has been verified. There is no safe alternative to eye cosmetics, but isn't your vision more valuable than a temporary exotic effect?

Shampoos, Protein, and Dandruff

Shampoo manufacturers would certainly have the consumer believe that everyone has a problem with his or her hair. Some heads have too much oil, others are too dry, while still others are plagued by dandruff and limp, thin hair.

A current gimmick for shampoos is "protein," which, it is claimed, will thicken thin hair and give it body. This additive is not, in fact, real protein, but rather a protein hydrolysate, which is a solution of several amino acids found in protein. While amino acids have shown some success in repairing split ends of long hair, they are unnecessary for cleansing and actually hinder the process. Additional amino acids glued to the hair catch dust at a rate that makes shampooing necessary more often than if the product were not used.

Dandruff, oiliness, and flakiness of the scalp are normal conditions resulting from the body's elimination of wastes and sluffing off of dead skin. Most people who complain about the condition simply don't wash their hair often enough. If the weather is excessively hot, if we are under emotional stress, or if there is a lot of dirt in the air, the scalp responds by putting out an extra amount of sebum and perspiration. Occasionally, a hormonal imbalance will result in an excessive amount of dandruff, in which case a doctor should be consulted. But for common dandruff, washing your hair daily with baby shampoo, or any other shampoo that is pH buffered to match the skin's acid mantle, will help.

The market shelf is also filled with an endless variety of cream rinses and conditioners. In general, they are of no more value to the hair or its appearance than the small amounts of milk or eggs they may contain.

Hair Sprays

Once strictly a woman's beauty aid, hair sprays are now equally popular with men. Compounded of a synthetic resin (polyvinylpyrrolidone—PVP) dissolved in water and alcohol, hair sprays are propelled by environmentally dangerous fluorocarbons that threaten to destroy the ozone layer of the stratosphere with their cumulative effect.

Sprayed onto the head in a fine mist, this deadly mixture is breathed into the lungs and lodges in the respiratory tract. Until recently, many doctors were puzzled by lung abnormalities that showed up in X-rays of patients complaining of breathing difficulties. Each person was found to have used hair sprays daily in a small, unventilated bathroom, and when they stopped using the products, the conditions cleared up dramatically. The illness has come to be known as thesaurosis, or "storage disease," and its steady increase has become alarming.[5]

Other complaints to the F.D.A. concerning hair sprays include hair loss, ringing in the ear, headaches, and several cases of hair spontaneously igniting when a cigarette was lit. By substituting a liquid or gel setting lotion you can effectively keep your hair in place.

Hair Coloring

Changing the color of the hair involves a massive assault on the hair's natural chemical structure. If the hair is lightened, the existing color must first be stripped away from the hair shaft with a caustic solution of hydrogen peroxide and ammonia. Next, a coal-tar dye (also known as an aniline dye) penetrates the hair shaft, swelling it up considerably and making it brittle. Once begun, the process must be repeated on the new growth about every four weeks, which is barely enough time for the scalp to recover from this caustic trauma.

Medical investigators have found many poisonings each year directly related to exposure of the hair follicles to coal-tar dyes, whose carcinogenic (cancer-producing) effects have long been recognized. In addition, these poisons seriously damage the liver and kidneys as the body struggles to expel their residues.

Gray hair is anathema in our society, and many men who would not dream of dying their hair pour on or pass through with treated combs equally harmful substances to attain a "natural" brown color. Called "color restorers," these dyes progressively deposit a dark, flat shade on the hair after several days' application. They also deposit a fine metal dust of pigments—usually lead—that have been found to cause severe irritation of the nasal passages, throat, and lungs. In some cases the irritation has been so severe that the tissues of the respiratory tract swelled and blocked breathing.

Graying hair—which is beautiful in its own right—can be brightened and softened by any number of non-irritating blueing rinses available on the market: regular rinses with camomile tea will add highlights to fair hair, and a bath of steeped henna leaves will brighten auburn hair.

Permanent Waving

In order to reshape the hair, the protein molecules must be unbound by a highly irritating chemical called ammonium thioglycolate. This chemical and others used in conjunction have frequently caused serious burns to the scalp and face, and have resulted in the permanent loss of hair. Occasionally, the setting lotions are splashed or dripped into the eyes where they cause irreparable damage. The *Journal of the American Medical Association* reports a case of a 53-year-old woman who accidentally got permanent wave-set lotion in her ear. The corrosive action pierced the woman's eardrum and caused permanent partial deafness.[8] Swelling of the legs and feet, and damage to the mucous membranes have also been reported.

At best, permanent waving will eventually result in brittle hair that is dull and lusterless. Set your hair with water or a neutral setting agent such as milk and you will avoid the potentially irreversible effects caused by these caustic and extreme treatments.

Hair straighteners could be damaging to the condition of the hair. They tend to break and damage the hair shaft, especially when used regularly.

Depilatories

Western women have been conditioned to consider excess body hair repulsive, and an estimated 98 percent of American women between the ages of 14 and 44 remove it.[9] The most commonly accepted method is to shave it off the arms, legs, and armpits. In recent years, however, many women have turned to aerosol, cream, or wax based chemical depilatories that dissolve the hair so that it can be washed off in the shower.

The active ingredient in depilatories is calcium thioglycolate, a stronger cousin of the permanent waving chemical and even more hazardous. Because hair is composed of tissue similar to skin, depilitories, if left on too long, can eventually cause deep, third-degree burns that can result in hideous scarring. Depilatories should be considered potentially harmful, as one accidental spray in the mouth or eye, blending of cream on the lip, or actual tearing of the skin will destroy tissues.

Shaving with a safety razor is the safest solution, although nicks and scratches can result and lead to infection. The safest home method of removing body hair is the electric razor after first applying a "pre-shave" solution to protect the skin.

Nail Polishes and Nail Hardeners

Nail polishes contain cellulose nitrate, which is a safe substance obtained from the walls of plants, and butyl acetate, which is a highly toxic synthetic acid. Butyl acetate is a narcotic when inhaled in large concentrations and its fumes have been known to cause conjunctivitis, an inflammation of the eye. Moreover, it can cause nausea, splitting nails, as well as permanently stain nail beds black. The dyes used are not known to be harmful.

Cuticle removers and nail polish removers are not dangerous to the nails or skin. However, it is wise to avoid inhaling their fumes.

Nail hardeners are sold to keep the nails from breaking and chipping; their primary ingredient is formaldehyde, the same chemical that is used to preserve dead bodies. Formaldehyde has been known to cause atrophied cuticles, lost nails, and damaged nerves at the fingertips.

Brittle, splitting, separated nails are often caused by the physical abuse of hard work and harsh household detergents. Wear gloves when you do housework; keep your hands out of extremely hot water and your nails will improve noticeably.

Dentifrices and Mouthwashes

The high incidence of tooth decay in this country is primarily due to a diet rich in starches, sugars, and sweet desserts, and inadequate in proteins, vitamins, and minerals, especially calcium. All those peanut butter and jelly sandwiches, candy, cookies, and soft drinks adhere to the teeth where they form a perfect medium for bacteria. The bacteria multiplies into millions and produces an acid that dissolves the enamel and dentine of which teeth are made.

Toothpaste can help prevent tooth decay, it is true, but the cleanser used is less important than proper technique. Toothpaste and even baking soda neutralize mouth acid, but the key is in removing food particles that may build up between teeth. Brushing should take place in a gentle up-and-down motion, paying close attention to the area around the gum line, and should last at least five minutes. The operation must be repeated after *any* sugar or starch is eaten, to be followed by a thorough flossing with unwaxed dental floss.

In the 1960s, toothpastes appeared on the market claiming "super-whitening ability." There is no way to whiten teeth other than cleaning the debris of food from them, but there are ways of polishing the enamel so that they reflect more light and seem brighter. Stay away from any dentifrices that make such claims, for studies show that they may contain abrasives that wear away the enamel, or peroxide bleaches that break it down.

Mouthwashes claim to be oral deodorants, but in fact they do nothing for bad breath except mask it with a flavoring oil. Unpleasant breath comes from decaying teeth, from food left in the interstices of the teeth, or from a throat, lung, or stomach condition. If it persists, a doctor should be consulted. As for claims that mouthwashes destroy germs that cause odor, the National Academy of Sciences—National Research Council examined the evidence and came to the following conclusion: "There is no convincing evidence that any medicated mouthwash, used as part of a daily hygiene regimen, has therapeutic effects over a physiologic saline solution or even water."

Notes

1.   U.S. Department of Commerce, 1970.

2. Winter, Ruth, *A Consumer's Dictionary of Cosmetic Ingredients*, New York, Crown Publishers, Inc., 1974.
3. Wells, F.V., and Lubowe, Irwin, *Cosmetics and the Skin*, New York, Van Nostrand, 1964.
4. U.S. Food and Drug Administration, 1974.
5. Consumer Reports, *Aerosols: The Medical Cost of Convenience*, May, 1974.
6. "Paraben Allergy—A Cause of Intractable Dermatitis," *Journal of the American Medical Association*, Vol. 204, No. 10, June 3, 1968.
7. Brodeur, Paul, *Asbestos and Enzymes*, New York, Ballantine Books, 1972.
8. Stabile, Toni, *Cosmetics The Great American Skin Game*, Ballantine, New York, 1973.
9. "Chemical Depilatories," *Consumers' Research Magazine*, January, 1974.

Tobacco, Alcohol, and Caffeine—Legal Narcotics

What sane person would condone the legalized sale of heroin or cocaine, for we are all aware of the deadly consequences of taking these physically and psychologically addictive drugs. They are, of course, prohibited throughout the civilized world on the grounds that they are dangerous and serve no useful medical function.

Yet, at the same time, we permit and encourage the sale and use of three similar poisons that are equally damaging to the body and mind. Tobacco, alcohol, and caffeine are nothing more than legally available, socially approved addictive drugs. Many of us smoke, drink too many alcoholic beverages, and gulp down cups of coffee and bottles of cola drinks, needlessly inflicting deterioration and disease upon ourselves, and thereby shortening our lifespans.

Because the toxic effects of tobacco, alcohol, and caffeine accumulate slowly, many users are lulled into a false sense of security since no immediate harm is apparent. Sadly, the realization of the deadly potential of these drugs usually comes only after the appearance of symptoms of cancer, emphysema, heart disease, or psychosis, many years later. By then it is usually too late to reverse the damage.

Advertisers do their best to lure the unwary victim down the primrose path of the addictive products they push. To be well liked and successful, Madison Avenue tells us, we must drink and smoke the way the handsome, healthy people do in their ads. Making friends is most successfully accomplished over cigarettes and a cup of steaming coffee or an iced glass of cola. Advertisements and commercials strive to create the impression that cigarettes, liquor, and caffeine drinks are somehow beneficial to the consumer; the grim reality is hidden behind a smokescreen of seductive, glowing words and images. Let's clear away the deceptions and take a close look at the truth.

Cigarettes and Tobacco

In 1964, *The Report of the Surgeon General's Advisory Committee on Smoking and Health* once and for all laid to rest the myth that tobacco smoking is a harmless and at times beneficial habit. The medical establishment had long suspected that smoking was a direct cause of cancer, lung disease, heart and arterial failure, and a number of other illnesses, but most previous studies had been considered inconclusive. The fact that an agency of the United States Government had publicly declared that smoking was a menace to public health was front-page news, and the *Report* quickly became the story of the year.

Alarmed, hundreds of thousands of cigarette smokers immediately gave up the habit, and the number of cigarettes sold dropped for the first time since they were rationed during World War II. The United States Public Health Service and the American Cancer Society began an all-out campaign to publicize the dangers of tobacco smoking, and offered suggestions to those who were trying to stop. Cigarette advertising was eventually banned on television and radio, and cigarette packages were required to carry the warning: "May Be Hazardous to Health." Later the wording was strengthened to read: "Smoking Is Dangerous to Your Health."

A month after the Surgeon General's *Report* was issued, retailers began complaining that cigarette sales had decreased by as much as 30 percent, and that their livelihoods were being threatened by the anti-smoking campaign. Their fears proved to be unfounded, however, for several months later cigarette consumption climbed back to its pre-1964 levels. Having spent millions of dollars trying to educate people with brochures, billboards, newspaper and magazine advertisements, radio and television commercials, and lectures, the Public Health Service conducted a survey to find out what had gone wrong. In interviews with a cross-section of current male smokers, the P.H.S. was surprised to learn the following:[1]

71.3 percent agreed that smoking is harmful to health;

59.5 percent hoped that their children would never smoke;

57.7 percent agreed that cigarette smoking is a cause of lung cancer;

54.6 percent agreed that smoking is a dirty habit;

44.9 percent agreed that there is something morally wrong with smoking cigarettes.

Clearly, the anti-smoking crusaders had gotten their message across to the people they had hoped to reach, but with little effect on actual smoking. That the spirit was willing, but the flesh weak, was all too apparent. What had happened was that in their haste to publicize the evils of tobacco, the P.H.S. had not given enough consideration (and still does not do so) to the addictive nature of nicotine. Smokers are hooked as surely as is any heroin addict; giving up cigarettes creates painful withdrawal symptoms and a craving for them that many people are unable to overcome. It is hardly a question of simple willpower. "If it were not for the nicotine in the tobacco smoke," writes Dr. M.A. Hamilton Russell, a British expert on drug addiction, "people would be little more inclined to smoke cigarettes than they are to blow bubbles or light sparklers."[2]

This physical dependence was first proven by Dr. Lennox Johnston, who conducted a series of experiments with smokers in his London research laboratories in 1942. When volunteers were injected with dosages of nicotine equivalent to the amount found in an average cigarette, they discovered that their desire for cigarettes suddenly vanished. As long as the injections were continued throughout the day, not one of the subjects wanted to smoke. When the injections were withdrawn, the craving returned. Dr. Johnston came to the conclusion that "smoking tobacco is essentially a means of administering nicotine, just as smoking opium is a means of administering morphine."[3]

Further proof has been offered by numerous studies, the most important of which was sponsored by the University of Michigan Medical School in 1967. Volunteer smokers were placed inside soundproof isolation booths for 6 hours a day for 15 consecutive days, during which time intravenous needles were inserted into their arms. The subjects were permitted to read and smoke; they were not told that the experiment had anything to do with the effects of tobacco. On certain days, a nicotine solution was fed through the needles, and on others a plain salt solution was substituted. The results were invariable: on nicotine-fed days the volunteers smoked significantly less than usual or not at all.

Researchers discovered that nicotine is carried by the bloodstream to the brain within a minute or two of smoking, and eliminated about half an hour later, when the craving returns. That nicotine is a deadly poison has long been known to scientists and farmers, who continue to use a concentrated spray of the chemical extracted from tobacco leaves as a powerful insecticide. In humans, nicotine constricts the blood vessels, thereby decreasing blood circulation to the skin and vital organs. Long-term smokers tend to look much older than non-smokers—a result of the contraction of the capillaries on the skin's surface, which prevents the absorption of tissue-building nutrients. Furthermore, smokers afflicted with arterial hardening and cholesterol deposits suffer a significantly higher number of heart attacks than non-smokers. Damaged blood vessels give way sooner when shrivelled by nicotine.

Until the early 1900s, tobacco was usually chewed, inhaled as snuff, or smoked in cigars and pipes without being inhaled. In other words, nicotine was absorbed into the bloodstream through the membranes of the mouth, nose, and bronchial passages—not through the lungs. The invention of cigarette paper and automatic rolling machinery changed all that, and soon tobacco users were puffing away at convenient, white-wrapped sticks of tobacco that introduced new toxins deep into their bodies. Known collectively as "tar," these toxins are by-products created from the combustion of paper, tobacco, and chemicals used in tobacco processing.

In healthy individuals, the lungs are elastic, rhythmic organs that exchange oxygen in the air for carbon dioxide and other waste gases emitted by cellular metabolism. If separated and laid end to end, the clusters of tiny, bubble-like air sacs (alveoli) that effect this process would fill a space equivalent to the length and width of a swimming pool.

The inhalation of cigarette tar causes the thin alveoli

membranes to break down; thus the lungs lose their natural resilience. Still trying to perform their function of exchanging gases with the blood, these air sacs form larger sacs which tend to trap carbon dioxide. Pressure builds inside the lungs, and it becomes difficult to breathe. In the constant struggle for air, the bronchial tubes and windpipe also lose their elasticity, and the alveoli begin to resemble balloons that have been blown up too long—when deflated they simply wrinkle instead of returning to their original size. This abnormality is known as emphysema, and there exists no cure for it.

The first symptoms of emphysema are usually breathlessness upon minor exertion, spells of coughing that leave one weak, and an excess of mucus in the respiratory tract that makes a bubbly sound as one breathes. It may surprise the reader to learn that many non-smoking city dwellers unwittingly suffer from this disease from breathing polluted air. In May, 1967, the *National Tuberculosis and Health Association Bulletin* reported the shocking news that the number of deaths caused by emphysema and bronchitis is doubling every five years. Obviously, coping with environmental irritants is bad enough; smoking cigarettes in addition compounds the problem.

The most lethal byproduct inhaled from burning tobacco is benzo(a)pyrene, a carcinogenic chemical that is also emitted from automobile exhaust pipes and factory smoke stacks. In numerous tests conducted at universities in the United States and England, benzo(a)pyrene has been applied to the respiratory tracts of laboratory animals, and has usually resulted in malignant tumors. The American Cancer Society identifies this smoke-borne substance as a prime cause of lung cancer.

Rather than attacking the small alveoli sacs, lung cancer begins on the walls of main bronchial air passages, whose cells begin to grow erratically after years of constant irritation. If the tumors grow inwards, they may cause collapse and infection of the lung, and the patient will cough up blood-stained sputum. These symptoms are considered fortunate, since they are an early warning, and the disease can sometimes be checked by the surgical removal of the infected portions of the lungs. If the tumors grow outwards into the surrounding tissues, there are often no early symptoms, and the disease silently spreads throughout the body, leaving little hope of cure.

The leading killer among all forms of cancers, lung cancer currently claims between 250,000 and 300,000 victims annually, which represents an astonishing increase of over 100 percent in the last 24 years.[4] Not surprisingly, the rise closely parallels a rapid increase in the number of cigarettes smoked over a similar period: in 1950, Americans consumed 392 billion cigarettes; by 1974 the figure had climbed to over 650 billion.[5]

Not included in available cancer statistics are additional malignancies suffered by cigarette smokers. The chemical irritants absorbed into the blood (including nicotine) are excreted almost unchanged in the urine, and they foster the development of cancer of the kidneys, prostate glands, and bladder. Even smokers of pipes and cigars are not immune. Although their incidence of lung cancer is lower, they run the risk of developing cancerous sores in the mucous membranes of the mouth and nose, due to the much stronger tobacco contained in these "smokes." Furthermore, many pipe tobaccos and cigars are sugar-cured, which further irritates the body.

As was pointed out earlier in this chapter, most smokers have been made aware of the harmful effects of tobacco. Yet few have been able to break the habit. Seventy million Americans continue to smoke and to run the risk of contracting the nastiest and most incurable diseases known to humans, thereby shortening their lives by several years. The American Cancer Society estimates that 75 percent of lung cancer deaths could be avoided if only people would stop smoking.

Some help has been given by new local and federal ordinances which forbid smoking in many public places, thus forcing smokers to cut down. Now, at least those of us who do not smoke no longer have to inhale the carcinogens produced by those who do. The most effective solution, of course, would be for the government to impose a ban on cigarette manufacture, and finally recognize cigarettes and tobacco as the dangerous substances they are.

But it is still possible to kick the habit! Since nicotine is a narcotic, abrupt withdrawal is rarely successful, as the craving usually quickly overcomes the desire to quit. Many experts advise that the amount of cigarettes smoked should be lessened gradually. Chew sugarless gum in place of every other cigarette. When the desire for a cigarette becomes overwhelming, draw in a deep breath; most ex-smokers have found this technique to be highly effective. Eventually the need for nicotine will abate several days or weeks later (depending upon your tolerance level), and willpower can take over.

The Alcohol Drugs

Alcohol addiction is second only to nicotine addiction as a prime cause of body pollution. In the United States, and many countries of Eastern and Western Europe, alcoholism has reached epidemic proportions, and the increase continues to spiral. The number of alcoholics in the United States alone has been estimated to be between five and seven million, and that figure does not include so-called "problem drinkers."[6] Tens of millions of work hours are lost yearly by alcoholics, and, according to the F.B.I., more than half of the nation's annual 50,000 fatal automobile accidents are caused by drunken drivers. The misery and tragedy that alcoholics inflict upon themselves and their families are beyond assessment.

Taken occasionally and in moderation, a highball, a bottle of beer, or a glass of wine will probably not harm you any more than a piece of candy. However, because alcohol is an addictive drug, most people find it difficult to limit themselves to one or two drinks per week or month. Before long, many social drinkers find themselves imbibing more and more, and progressing from problem drinker to the ranks of the estimated 10 to 12 percent of all drinkers who become alcoholics.[7] The social drinker is gambling with his or her health and peace of mind, for no user can predict with certainty whether he or she will become addicted to alcohol.

Inexplicably, the F.D.A. views alcohol as a nondrug, and permits its open sale in hundreds of thousands of retail stores, restaurants, and bars. At the same time, barbiturates—which produce the same mind-altering effects as alcohol—are legally available only by prescription. In addition, their use has been severely restricted under the Drug Abuse Control Amendment of 1965. Meanwhile, producers of alcoholic beverages spend over $250,000,000 yearly on advertising that encourages consumers to ingest a chemical that causes even more physiological harm and social disintegration than a drug that has been suppressed.

Like barbiturates, alcohol depresses the central nervous system to produce the desired effects of relaxation and moderate euphoria. Similar to a general anesthetic, alcohol numbs the control centers of the brain, releasing normal inhibitions and increasing irrational behavior. One too many drinks has turned many an otherwise sensible man or woman into "the life of the party." Alcohol also decreases alertness and motor ability, as evidenced by the staggering drunk, who with face flushed and distorted tries to keep from slurring his or her words. Eventually, the alcohol "high" turns into depression, the brain shuts off, and a dead sleep arrives. In the morning there is the inevitable hangover.

The nauseated, headachy, washed-out, tired feeling that constitutes a hangover is a signal that the liver is rebelling from a massive assault of alcohol. Excessive thirst is a reminder that alcohol is also a diuretic, which irritates the kidneys and causes them to excrete more water than is taken into the body. Both conditions are evident in the hangover sufferer's appearance. The puffy, gray look occurs because the liver—the major cleansing organ of the body—is unable to completely filter toxins from the bloodstream and is being suffocated by them. The dry, mottled appearance of the skin means that the tissues have been leached of water, and nutrients are unable to be absorbed. Ugly red networks of burst capillaries on the face and body attest to alcohol's strong dilating effect on blood vessels.

Constant drinking slowly poisons the liver cells, which are eventually killed off. Striving to correct the damage, the remaining liver cells generate new ones while the dead cells turn into scar tissue. The regenerated cells form into knob-like clusters separated by thick bands of hard scar tissue, and the result is cirrhosis of the liver. Such a liver is grossly misshapen and enlarged, and often distends the abdomen to form the familiar "beer belly" or "drinker's gut." Fortunately for the reformed drinker, the liver is a truly remarkable organ that has the power to rebuild itself even after up to 80 percent of it has been destroyed.

Sorrowfully, many alcoholics are so badly addicted that they fail to heed the warning signs of jaundice (yellowing of the whites of the eyes and skin) and bloody vomiting, and they proceed to destroy what is left of their livers. Death usually comes in the form of a massive hemorrhage.

Often, cirrhosis develops into cancerous cells rather than healthy new growth, in which case there is little hope for recovery. The lining of the alimentary canal is easily irritated by alcohol, and chronic abuse may produce cancer anywhere along the line, or stomach ulcers, before symptoms of cirrhosis are manifest. The warming sensation experienced when a drink is swallowed is caused by alcohol burning esophageal tissue, and there is nothing beneficial about it.

Cirrhosis is due in part to the fact that heavy drinkers fail to eat properly. They obtain much of their daily quota of calories from alcohol, which fuels the body without providing protein, fat, vitamins, or minerals, and they consequently suffer symptoms of malnutrition very similar to those endured by people starving in underdeveloped countries. Among the most serious nutritional diseases associated with alcoholism are neuritis, a painful inflammation of the nerves; beriberi, a vitamin B-1 (thiamine) deficiency that brings about an enlarged heart, excessive bowel activity, gastric disturbances, and general decline; and pellagra, a vitamin B-3 (niacin) deficiency disease that results in weakness to the point of helplessness, rashes on the hands, arms, and face, diarrhea, and often emotional breakdown. Being malnourished, drinkers have a lowered resistance to infectious diseases, and are inclined to contract lobar pneumonia, which has a higher death rate among drinkers than nondrinkers.

Moreover, alcohol addiction is as destructive to the mind as it is to the body. Years of constant drinking destroy brain cells, resulting in loss of memory, hallucinations, and paranoid delusions. Delerium tremens ("D.T.'s"), in which the alcoholic thinks he or she sees pink elephants or feels insects crawling under the skin, is another sign of imminent mental deterioration. The problem has become so severe in recent years that a Presidential commission was formed to investigate alcohol's impact on the public. The commission uncovered the following eye-opening data: More than 22 percent of all men admitted to state mental hospitals for the first time in 1964 were diagnosed as alcoholics (the figure for women was 5.6 percent; and approximately 29 percent of male patients in general hospital wards were problem drinkers.[8]

Because alcohol releases the social inhibition controls and lessens alertness and coordination, drinking has become one of the biggest law-enforcement problems in the United States today. This cost is also borne by the non-drinking taxpayer, who must daily face death or permanent disability when driving on highways infested with drunken drivers. According to the Presidential commission's report, "In 1965, out of close to five million arrests in the United States for all offenses, over 1,535,000 were for public drunkenness (31 percent). In addition, there were over 250,000 arrests for driving while intoxicated. Another 490,000 individuals were charged with disorderly conduct, which some communities use in lieu of the public drunkenness charge. Thus, at least 40 percent of all arrests are for being drunk in a public place, or being under the influence while driving..."

Crimes of violence are more likely to occur when people are under the influence of alcohol, the Commission points out, citing a 1958 study of homicides committed in Philadelphia. In 44 percent of all murders, autopsies showed that both the killer and victim had been intoxicated. A similar 1953 study in Texas revealed that 28.5 percent of all murders there took place in bars, saloons, cocktail lounges, and other places where liquor was served. Children very often become the direct victims of parents with a drinking problem, and end up as statistics of the "battered child syndrome." Disturbed and unable to face the ever-increasing problems incurred by their alcoholism, such parents get drunk and take out their frustrations on their children, beating them black and blue and sending them to the hospital—or to the cemetary.

Like drug addicts, alcoholics are sick human beings who have become enslaved to a chemical that is slowly poisoning their minds and bodies while bringing grief and tragedy to those around them. Most of us believe alcoholics to be obvious skid row types who have given up family, friends, business, sex, and all the other activities that enrich our lives, in their quest for yet another drink. But they are only part of the problem. For every unshaven, toothless wreck of a man, there are dozens of neatly dressed men and women, many of whom drink secretly, and are addicted to alcohol. Although they appear to function normally they harm themselves nearly as badly. They are the people who drive while intoxicated, whose doctors thoughtfully admit them to private hospitals when they become ill, and cover up the fact that they are alcoholics.

A cure can only be effected when the person admits that he or she is an alcoholic. Psychotherapy may be undertaken to help to deal with the psychological and social problems. The chemical imbalance that alcohol has created in the body—which is the source of addictive craving—can be corrected by megavitamin therapy administered by a doctor. Antabuse, a medicine that causes distress when the patient takes a drink, may also be helpful. The best help by far is available from Alcoholics Anonymous, an organization of individuals who have conquered their drinking problems and are trying to help others; they also provide a list of doctors and psychiatrists who are experts in this field.

Most social drinkers can avoid becoming alcoholics by drinking only on special occasions, and by swallowing copious amounts of water so as to dilute the alcohol. If you must drink, stick with beer and wine and always take some food while drinking. A daily multivitamin capsule is essential, along with a B-complex supplement to replace the amount destroyed by alcohol. Above all, never drink to escape an emotional problem. Seek the advice of a friend, or professional help.

The Caffeine Habit

Why is that morning cup of coffee so delicious and satisfying? You guessed it—the caffeine in coffee is addictive as well as stimulating, and it satisfies the craving built up by the body during the night's sleep. Similar in effect to amphetamines and cocaine, caffeine stimulates the cerebral cortex and speeds up the thinking processes while defeating drowsiness and fatigue. In other words, it helps you to wake up and keeps you awake.

Contained in tea, cocoa, and cola drinks, as well as coffee, caffeine provides false energy for a brief period, and increases simple motor skills such as typing and driving. But the chemical action of caffeine is no substitute for the fuel provided by a good breakfast, a fact all too apparent in the mid-morning energy lag experienced by workers who begin the day with only coffee.

Laced with cream and sugar, a cup of coffee is a metabolic insult that places a concentrated burden on the pancreas, which may well result in chronic hypoglycemia—low blood sugar. Immediately absorbed into the bloodstream, the sugar sounds an alarm bell for the pancreas to produce insulin to turn it into energy producing glucose. Caffeine stimulates this effect, as does the nicotine in a cigarette. In an hour or so the excess insulin has burned up the sugar in the blood, and—after an initial burst of morning energy—the caffeine-drinker is left with the tired, listless feeling that is symptomatic of hypoglycemia.

The morning refreshment wagon brings another coffee and an insulin-producing, heavily sugared sweet roll, and the person is able to make it through to lunch. The afternoon break will undoubtedly add more caffeine and sugar into the worker's disrupted system in the form of an iced cola drink. Eventually, the stimulating effects of caffeine can no longer overcome the debilitating effects of low blood sugar, and the drug merely makes the user dizzy, nervous, and uncoordinated.

However, if more than four or five cups of coffee are consumed daily, the same symptoms can appear in a person with a normal blood-sugar level. A now-classic report of a case of caffeine abuse was recorded in the *New England Journal of Medicine* in 1936. On the advice of an intern at the hospital where she worked, a nurse began taking a grain and a half of caffeine citrate three times a day to overcome fatigue produced by long hours of strenuous duty. Such tablets were (and are) widely available at most drugstores under such trade names as *NoDoz*, and unfortunately require no medical supervision.

Ingesting the equivalent of nine cups of coffee daily, as well as her normal four cups, the nurse first began to suffer one of the most common symptoms of too much coffee:

she couldn't sleep at night. Sleeping tablets were needed to put her to sleep and more caffeine tablets to wake her up. The addiction to caffeine slowly increased her tolerance level to the point where she was swallowing five, ten, and twenty of the capsules in order to function properly.

Finally, trying to pep herself up for a staff party, the nurse swallowed the contents of the box—forty tablets— and the dam burst. "She became confused, disoriented, excited, restless, and violent," the report states. "[She] shouted and screamed and began to throw things about her room ... Finally she collapsed and was removed to a general hospital." Unaware of her dependence on caffeine, the hospital staff diagnosed her case as "psycho-neurosis, anxiety type, with a hysterical episode," and she was transferred to a psychiatric ward, where she was strapped to a bed to protect her from harming herself. Eventually, the poison was eliminated by her body, and the nurse was able to recount what had happened. Taken off caffeine tablets and coffee, she recovered completely and returned to work a few weeks later.[9]

As you can see, the coffee-caffeine habit can become a serious one, albeit not as acute as this classic case. Most heavy caffeine imbibers run the risk of slow-acting, permanent damage from years of drinking their favorite steaming hot brews and cold soft drinks. The kidneys are constantly irritated from caffeine's diuretic effect, which can lead to infection and subsequent kidney failure. The stomach lining is also irritated, and gastric ulcers are a common complaint of people who drink too much of this stimulant.

More alarming is caffeine's relationship to heart disease. In 1973, investigators at several Boston hospitals compiled a study of the coffee drinking habits of all patients admitted during the previous years. They found that drinkers of more than five cups of coffee a day had twice the incidence of coronary heart attacks of non-drinkers.[10] Researchers in Pennsylvania and Canada have concurrently found proof that a high level of caffeine in the blood increases the amount of cholesterol circulating in the arteries.[11] It has already been established that caffeine stimulation promotes an irregular heartbeat, increases blood pressure, and plays a part in inducing diabetes.

Until this year, few people suspected that caffeine consumption could also be responsible for some of the 250,000 malformed and retarded babies that are born annually. Now, the Center for Science in the Public Interest has found a definite correlation between the two, as a result of lengthy experiments conducted on test animals. Fed the amount of caffeine present in eleven cups of coffee, many of the animals gave birth to offspring with deformed heads, missing fingers and toes, and cleft palates.

"Although the animal and human studies do not prove conclusively that caffeine causes birth defects, miscarriages or infertility in humans," the Center report cautions, "the evidence is suggestive enough to require a public education campaign and more extensive research." The H.E.W. advises pregnant women, particularly those in their first three months of pregnancy, to restrict their intake of coffee, tea, cola drinks, and aspirin compounds, many of which contain caffeine.[12]

While most parents are wise enough to forbid their children to drink coffee, they nevertheless expose them to the same risks by allowing them to drink cola soft drinks, cocoa drinks, and tea. The resultant overstimulation often produces symptoms of nervous restlessness that many doctors misdiagnose as hyperkineticism, which is an entirely different condition. Needless medications are often administered to correct a disease that isn't present, further polluting the body rather than cleansing it of the cause of the problem.

Tea, cocoa, and cola drinks contain about half as much caffeine as coffee, but of the three, only weak tea can be recommended. Cocoa contains an additional stimulant, theobromine, which brings the total level of addictive chemicals up to that of coffee. Cola drinks, of course, are loaded with equally addictive white sugar and chemical additives. Decaffeinated coffees are no solution, either, in view of the residues of chemical solvents used to extract the caffeine. Furthermore, all of these beverages contain a number of chemical compounds as yet unidentified, which may also inter-react negatively with the body's normal processes.

Wean yourself from these addictive chemical concoctions—which encourage many to exacerbate the damage by picking up the habit of smoking. Switch to delicious, invigorating herb teas that even supermarkets are beginning to carry, or unsweetened fruit juices. Your taste buds will eventually come back to life after years of having been dulled by caffeine pollutants, and food will begin to taste good once again.

Notes

1. National Clearing House for Smoking and Health, U.S. Dept. of Health, Education and Welfare, *Use of Tobacco*, 1969.
2. Russell, M. A. Hamilton, "Cigarette Smoking; a Natural History of a Dependence Disorder," *British Journal of Medical Psychology*, (44), 1971.
3. Johnston, Lennox M., *Tobacco Smoking and Nicotine*, Lancet, December 19, 1942.

4. The American Cancer Society, 1975.
5. U.S. Dept. of Agriculture, Economic Research Service, *Tobacco Situation,* 1975.
6. National Institute on Alcohol Abuse and Alcohol, 1975.
7. *Ibid.*
8. The President's Commission on Law Enforcement and Administration of Justice, Appendix I, *Task Force Report: Drunkenness,* 1967.
9. McManamy, Margaret C., and Schube, Purcell G., "Caffeine Intoxication," *The New England Journal of Medicine,* (215) 1936.
10. *Science Digest,* June, 1973.
11. *Consumer Bulletin Annual,* 1972.
12. *The New York Post* (The Associated Press), February 3, 1976, p. 67.

The Great American Medicine Show—"Pure Theatre"

Nowhere is our sad state of health more apparent than in the enormous quantity of medicines we use. According to popular belief, no one need suffer ill health these days because there is a pill or an injection to cure everything from headaches to heartaches. Even cancer and heart disease, we believe, will eventually be conquered by medical science.

Unfortunately, much of what we know about medicine is inspired by the publicity staffs of large drug manufacturers, whose prime consideration is to sell their products. Their business is peddling cures, not preventing disease and sickness. Although a wealth of evidence points to emotional stress, faulty nutrition, environmental pollutants and lack of exercise as the chief culprits in causing heart disease and cancer, for instance, drug merchants persist in spending millions searching for illusive if not totally nonexistent "cures."

The money would be better spent educating the public to avoid stress and unnatural foods contaminated with additives, and to get adequate rest and sleep. The body would thus be able to heal itself with its own remarkable recuperative powers, and a minor infection would be prevented from escalating into a debilitating disease. But the medicine hustlers turn a deaf ear to the time-proven adage: "An ounce of prevention is worth a pound of cure."

Instead, we are constantly bombarded with advertisements and drug-industry subsidized magazine articles that create alarms so that they can offer assurances. What we are not told is that many of these magical potions are useless, unnecessary—and even deadly! Prescription drugs kill 30,000 people annually, while thousands more suffer permanent damaging side effects, such as kidney damage, blindness, deafness, and anemia.[1] Statistics on fatalities caused by drugstore remedies are not available, but they are thought to be considerable.

Over-the-Counter Medications

Judging from advertisements, Americans suffer constantly from various aches and pains, coughs, runny noses, upset stomachs, and constipation. On television women relieve their husbands' misery in seconds by administering the "pain reliever doctors recommend most." Sprays open animated, stuffed-up nasal passages, magical elixers and capsules relieve coughs and post-nasal drip, tablets unclog sluggish bowels, and lozenges sweeten sour stomachs. You can buy these medications without a prescription much less a doctor's advice.

If left alone your body will heal itself anyway, since these ailments are usually more annoying than dangerous. Moreover, by tampering unnecessarily with your body's chemistry, you are exacerbating the illness with poisons that purport to cure it. Your body's self-healing mechanisms will have to overcome the effects of the drug before they can go to work on the malady.

An outgrowth of patent-medicine and snake-oil hustlers, manufacturers of these overpriced nostrums have

been called "the last of the robber barons" by a
Congressional drug investigating committee.[2] Their
greatest benefit is the work they provide for advertising
people.

Aspirin and Pain-Killers

Aspirin is the most widely accepted and commonly
used of all medicines. Over 27 million pounds of aspirin
are consumed yearly by Americans,[3] including its various
disguises as a remedy for colds, arthritis, and rheumatism.

Valuable as a reliever of headaches and other minor
pains, and as a fever reducer, aspirin is generally safe and
effective but only when used in moderation. That means
that no more than two tablets should be taken within eight
hours. Larger dosages may result in increased blood
pressure, rapid heartbeat, impairment of vision, and other
temporary side effects. Aspirin manufacturers fail to issue
these warnings, in fact are not required to do so, and
consequently thousands of users could be poisoned
annually, some of whom may suffer permanent circulatory, stomach, intestine, or liver damage.

The active ingredient in aspirin is salicylic acid, which
acts by raising the pain threshold of the brain. Since pain is
one of the body's danger signals, it is always wise to
consult a doctor if the discomfort persists. By alleviating
the symptom, you may be delaying the treatment of a
serious disorder.

Salicylic acid causes stomach irritation in many people,
a condition that some manufacturers claim to overcome
by "buffering" it with antacids. Such advertising claims
are pure Madison Avenue bunk, for all aspirin is alike
regardless of what has been added to it, and clinical tests
have shown the addition of bicarbonate of soda to be less
effective in soothing the stomach than a glass of milk or

two crackers taken with tablets. If you must use aspirin, always buy the cheapest "house" brands, since it is more likely they will be nonadulterated with buffering additives.

In recent years a new analgesic (pain-killer) alternative to aspirin has been marketed under different trade names. Costing as much as ten times more than aspirin, these tablets contain a compound named acetaminophen which has the advantage of not causing stomach upset. Other than that, the incidence of side effects has been found to be about the same as aspirin, making it a dubious improvement.

Remember that aspirin in large doses is a *poison*. Children should never be given dosages in excess of the recommendations listed on the bottle, nor should they be given flavored children's aspirin which they may later mistake for candy. Poisonings from children's flavored aspirin are a leading cause of death in the one-to-four age group.

Cold and Cough Remedies

There is no cure for the common cold. No medicine has yet been discovered or created that can destroy the 150-or-so viruses that invade the body and cause a stuffed nose, a cough, muscle aches, and a generally sick feeling. Purveyors of cold remedies choose to obscure this fact in high-powered advertising campaigns that claim to offer "prompt relief from suffering," when actually their products only further pollute the body.

Americans nevertheless buy these medications to the tune of $680 million a year,[4] mindless of the harm they are doing to themselves. Most of these drugs are known in the trade as "shotgun" formulas, meaning that they are compounds of three or four ingredients. The first is usually aspirin, the second a decongestant, the third an antihistamine, and the fourth may be ascorbic acid (vitamin C) or an antacid.

Decongestants reduce swelling of the membranes in the nose and upper respiratory tract and temporarily alleviate the stuffed-up feeling of a cold. But the membranes soon swell up again, the clogged feeling returns, so sufferers take another dose. The result is a condition called "rebound congestion," a swelling worse than the original one. Nasal sprays should be avoided, as they all contain decongestants and are breeding grounds for bacteria. The spray tip is contaminated every time the user puts it into his nose. Within a few days, millions of bacteria have multiplied in the fluid and are ready to repeat the infection.

Antihistamines are of equally dubious value. When an allergic reaction occurs, the body secretes histamines, defensive substances that can result in a runny nose and watery eyes. Antihistamines block this natural action only too well, and often the mucous membranes of the respiratory system dry out and a cough is brought on. At the very least antihistamines will make you drowsy, which is why they are also used as the active ingredient of over-the-counter sleeping tablets.

Coughing is a reflex response to an irritation anywhere in the lungs or the respiratory tract, and often accompanies a cold. More than 800 nonprescription cough medicines are available to the unwary, and most contain the same ingredients. Cough suppressants and expectorants are added to the mixture, further defeating the natural function of a cough, which is to bring up thick mucous secretions clogging the lungs and throat.

The only health-promoting ingredient in any of these preparations is vitamin C, which is included in such tiny amounts as to be virtually useless. Recently the F.D.A. concluded, after a three year study, that the majority of these remedies were not only ineffective but dangerous to health in their present combinations, and many are expected to be banned. A healthy, clean body will forestall most virus attacks, but if you nevertheless succumb, the best treatment for a cold is plenty of rest and plenty of water. It is also important to eliminate stress, which causes the body's autoimmune system to break down, making the person more susceptible to viruses and infections. A hard candy or a honey drop will do more to soothe your throat than the most expensive lozenge.

Laxatives

Television commercials to the contrary, "irregularity" is no reason for taking a laxative. There is so much variation in individual bowel habits from day to day that no one can accurately describe what constipation is. In a healthy person nature *always* takes care of the elimination of waste matter from the body. This usually happens at daily intervals, but it is not imperative that it does. Sometimes conditions such as dehydration cause the skipping of bowel movements for several days. It is nothing to worry about; nature will eventually cause the intestines to evacuate the accumulated waste material.

Some people regularly consume laxatives in the mistaken belief that waste matter in the intestine is poisonous and must be eliminated from the body as soon as possible. Waste matter is made up of undigested food, broken down blood cells, and other body debris. The toxins that occur in the body are taken into the intestine for elimination, but this is a one-way process—normally the toxins do not reenter the body.

True constipation can occur in people who are

bedridden or afflicted with severe emotional problems, in which case a doctor should be consulted. Hard stools and difficult passage are not symptoms of true constipation, but rather of a diet lacking in roughage. Add more fiber-rich foods to your meals, including bran, wheat germ, raw vegetables, and unpeeled fruit and you will find that bowel movements are both easier and more frequent. Laxatives only make matters worse by depleting the body of potassium, which causes muscle weakness and another harsh dose of laxative to get the bowels to work.

Antacids

Antacids are nearly as popular as aspirin and they are advertised as an antidote to that catchall malady, "indigestion." The word has no medical meaning, but is used to refer to many different irritations of the stomach. Nausea, gas, a bad cold, escape of acid into the esophagus ("heartburn"), and emotional distress can all produce similar symptoms.

Treating oneself with antacids is a very risky proposition, for stomach upsets may also be an indication of a more serious illness such as ulcers or even cancer. A chronic case of indigestion should always be brought to a doctor's attention and not self-diagnosed. Many people who thought they had eaten something that didn't agree with them never lived to find out that their gastric upset was actually the first symptom of a coronary heart attack.

For these reasons, antacids should be avoided and the cause of the indigestion determined by a doctor. In any event, they should never be taken over a long period of time, as the active ingredient of calcium carbonate may form stones in the kidneys and cause irreparable damage. A glass of warm milk is the safest antidote to occasional stomach distress.

Boric Acid

Once a common item in home medicine cabinets, boric acid powder is one of the most lethal poisons still freely available in drugstores. While it has mild germ-killing properties when dissolved in water and used as a lotion, boric acid powder can cause severe reactions if applied directly to cut or abraded skin. Inhaled, the powder is an acute irritant of the central nervous system, and in sufficient quantities it can cause death.

The American Medical Association has repeatedly urged the F.D.A. to ban this poisonous drug, but to no avail. The F.D.A. refuses to consider boric acid a poison because it is labeled "for external use only." This simply does not take into consideration the numerous cases of infant death caused by dusting with boric acid powder. In 1951, the *American Journal of Diseases of Children* reported a case in which a father put nine ounces of boric acid on the diapers of his nine-month-old daughter. She died 26 hours later of severe damages to the intestinal tract. The *Journal* article concluded with a warning: "Boric acid and sodium borate are sufficiently poisonous to cause severe symptoms and death when used in amounts commonly considered to be harmless."

Prescription Medicines: Rx—Debility, Deformity, and Death

During the last 30 years a revolution has taken place in medicine. A family of antibiotic "wonder" drugs was discovered and the public could now be cured of ailments that had formerly been disabling or fatal. Various hormones offered relief from glandular, skin, bone and blood disorders, as well as being used as the most effective means of birth control yet devised. Chemicals enabled us to control anxiety, they put us to sleep, and gave us energy when we awoke.

According to the drug industry, the millenium was at last at hand and only death remained to be conquered. Then doctors began to notice a frightening phenomenon—the cure was often worse than the disease! What had happened was that many miracle cures caused so many dangerous, even fatal, side effects that they spawned a new malady called *iatrogenic disease*, meaning those caused by medicines. In their haste to earn enormous profits, some irresponsible drug manufacturers foisted (and continue to do so) inadequately tested drugs on a gullible public, and tens of thousands of innocent people die or are crippled annually. Since many doctors rely heavily on drug therapy and prescribe too many drugs too often, it is imperative that the layman be alert to the dangers of the following medications.

Antibiotics

For all their undisputed benefit to mankind, antibiotics remain a double-edged sword. Penicillin and its 100-or-so cousins are remarkably effective cures for a variety of microbe-caused diseases, including pneumonia, scarlet fever, typhoid, tetanus, and venereal disease. Injected under the skin or taken orally, antibiotics swiftly attack and kill germs, funguses and some viruses, while leaving human tissue alone.

On the debit side, antibiotics also destroy many strains of bacteria normally present in the body. Ordinarily, these strains compete with each other and their number is kept low. When an antibiotic is administered, most of these bacteria perish along with the infection, leaving one or two strains to proliferate without competition. The bacteria explode in numbers, toxins overwhelm the body, and the result is one of the most feared hazards of antibiotic therapy—superinfection, which is fatal in 30 to 50 percent of cases.[5]

Many people are allergic to antibiotics, especially penicillin. Upon injection, the patient may go into anaphylactic shock, fall into a coma, and die. Adrenaline must be given within minutes as an antidote. Some antibiotics produce permanently debilitating side effects, a startling fact that has been soft-pedaled by the drug industry. Streptomycin may cause deafness and permanent loss of balance, and chloramphenicol may injure bone marrow, causing chronic anemia.

Heavily dosed over the years, bacteria have begun to build up a resistance to antibiotics, and it may not be long before dread "Andromeda-strain" microbes evolve that cannot be destroyed. In the spring of 1974, the Senate Health Subcommittee placed the blame squarely on the shoulders of drug industry promoters who push the drugs to physicians, who in turn overprescribe them. Most of us already have significant quantities of antibiotics coursing through our systems from eating the meat of livestock fed with them to promote growth. Constant exposure to antibiotics has given lethal microbes the means to mutate into new forms, while sensitizing us into allergic reactions.

Never use an antibiotic unless absolutely necessary to save your life, or to cure an otherwise hopeless condition. Always demand an allergy test, and ask your doctor about specific possible side effects.

Steroid Hormones

None of this group of widely publicized drugs cure disease. They act by reducing painful and aggravating symptoms in ways not yet clearly understood. Cortisone is the most-used steroid hormone and, like the others, it is a chemical that reproduces a secretion of the adrenal glands. Cortisone and its companion steroid, ACTH, can be remarkably effective in relieving symptoms of arthritis, rheumatism, skin and blood disorders, and even some types of cancer.

Undesirable and serious side effects often interfere with treatment, however, and steroids are not the panacea doctors once believed them to be. The most common problem is retention of water, leading to tissue swelling (edema) and high blood pressure. As additional hormones interfere with the body's natural metabolism, any number of seemingly unrelated conditions can result, including ulcers, osteoporosis (thinning of the bones), diabetes, and psychosis that mimic insanity. With long-term treatment, the adrenal glands shrink, and the body is unable to produce its own hormones once the drug is stopped. The result is that, in case of accident or an operation, the patient's own adrenal glands would probably not be able to stand up to the stress.

Tranquilizers

Thanks to a massive advertising campaign, tranquilizers are among the most prescribed drugs in the United States. Nearly as familiar as aspirin, tranquilizers are promoted as happiness pills that do everything from "restoring the zest for living" to "controlling anxiety, tension, irritability, and depression." In other words, it is unnecessary to deal with the stresses of life creatively. All you have to do is pop a pill to erase them.

Such claims may well be mere advertising puffery, judging from tests conducted in 1968 and 1969 by the Veteran's Administration and the National Institute of Mental Health. Half the patients who volunteered for the experiment were given the active drug, while the others were given placebos (dummy tablets) that looked identical. The tests were "double-blind," meaning that neither the doctors who administered the pills nor the patients knew which was which. The subjects were examined before and after the experiment so that the results could be judged objectively and not on the basis of what the volunteers thought they felt.

After six weeks, half the patients given tranquilizers showed a marked decline in "nervousness," tension, irritability, and so on. But so did nearly half the patients who were fed the placebos. "[Tranquilizers] generally come out as being a little better than a placebo, but not by any dramatic margin," concluded Dr. Jonathan O. Cole, head of N.I.M.H. (National Institute of Mental Health) drug research studies.

No evidence exists to support the extravagant claims of manufacturers of these useless drugs, which seem to work because the patient expects them to. The wholesale price of the market's leading tranquilizer is $1043 per troy ounce, making it 30 times the official price of gold. [6] A glass of warm milk or a cup of camomile tea will relax you as effectively.

Sleeping Pills

From time to time most of us experience difficulty in falling asleep or staying asleep. Usually, the cause is an especially tense day at home or at the office, or drinking coffee, tea, or cola drinks close to bedtime. Missing a few hours of sleep will not have an adverse effect on health, and the next night you will probably sleep quite well.

Not content to let well enough alone, pharmaceutical companies urge us to take sleeping tablets for a "deep, natural sleep." The claim is totally misleading for barbiturates and their synthetic substitutes simply knock us out. A barbiturate-induced sleep results from a massive depressant assault on the central nervous system and the spinal cord. Tolerance to the drug quickly builds and before long a larger dosage is needed to lull us into the unconsciousness that passes for sound sleep. Since the drug is a narcotic, addiction could occur.

Those addicted to sleeping tablets closely resemble alcoholics, in that they slur their words, seem mentally confused, and look generally unhealthy due to their tissues being poisoned. Drinking alcohol with barbiturates and other sleeping potions can easily result in death from suffocation, and is a not uncommon occurrence. Unable to resume a natural pattern of falling asleep, the chronic user often becomes impatient when the usual dose has failed to work, and he or she swallows several more tablets. In this intoxicated state, the person may forget how many doses have been taken, the tragic result often being unintentional suicide.

Those who try to kick the habit may suffer needlessly. The withdrawal symptoms are more severe than those of a heroin addict, and they include paranoid delusions, hallucinations, fever, violent tremors, and epileptic seizures.

Amphetamines

Known as "speed" by the drug culture, amphetamines were the most abused drug of the 1960s. Originally used against a rare but serious disease, narcolepsy (uncontrollable sleep), amphetamines are a prime example of the drug industry's high pressure sales techniques. In the 1950s, manufacturers began to promote the drug as an appetite-suppressant to counter the growing national problem of obesity. Soon, teen-aged students began to snatch Mom's diet pills because they seemed to keep them alert during all-night "cram" sessions before tests.

Indeed, the drug did stimulate alertness and wakefulness, as well as induce a euphoric high when injected. Demand grew throughout the 1960s and drug manufacturers increased production, seemingly oblivious to the fact that a large proportion of amphetamines were being diverted into the black market for thrill-seekers. Gradually, the corroded underside of the golden pill came to light. Emaciated, unkempt, and sickly speed freaks were a familiar sight in youth ghettos across the country, attesting to the mind and body destroying power of amphetamines.

Although not physically addictive, amphetamines encourage psychic dependency. Using them to obtain more energy only depletes the body of essential nutrients, and you will pay later in the form of illness or a psychotic mental state. Unlike foods, amphetamines offer no energy; they draw from existing reserves. Withdrawal usually results in severe depression that drives many back to the drug or to suicide, complete exhaustion, paranoia, aggressiveness, and brain damage. Deaths from heart failure have occurred among athletes who mistakenly tried to increase their endurance by using amphetamines.

Years late, as always, the F.D.A. finally imposed severe restrictions on the manufacture and distribution of amphetamines in 1972, effectively drying up the black market. The drug is, however, still available on prescription and there are many doctors still willing to prescribe them indiscriminately.

The Pill and Estrogen Hormones

Oral contraceptives make pregnancy impossible by inhibiting the release of new eggs (ovulation). A combination of two female hormones, progesterone and estrogen, the pill induces a state similar to pregnancy, when it is impossible to be re-impregnated. For this reason, morning sickness, nausea, and swelling of the breasts often accompany the contraceptive effect.

More important, the pill can also cause blood clots in the legs, high blood pressure, vaginal infections, fluid retention, and weight gain. Doctors have long suspected that the pill causes circulatory problems that could prove fatal, but it wasn't until recently that statistics offered irrefutable proof. A recent study undertaken by Planned Parenthood revealed that up to age 30, oral contraceptives are relatively safe, but that risk increases substantially with age. Between the ages of 30 and 40, complications occur often, and over age 40, use of the pill presents a definite hazard to health.[7]

Statistics compiled from the United States and Europe show that 25 per 100,000 women over 40 die yearly from blood clots, strokes, and heart attacks, all of which are directly a result of taking oral contraceptives. Obesity complicates the risk, as does smoking, diabetes, and high blood pressure. Women in this age group are advised to use safe contraceptive methods, such as condoms and diaphragms.

A question that many women ask is, "Does the pill cause cancer?" All the evidence is not yet in, but preliminary studies suggest that the answer is "possibly." Most of the side effects are attributed to a known cancer-causing substance—estrogen hormones. Post-menopausal women are often administered estrogens in order to restore a more youthful appearance and aid discomfort. Researchers at the University of Washington report that the incidence of uterine cancer is five times greater among women who took the hormone than among those who didn't. Also, the heavy fat content of Western diets may overstimulate normal hormone activity, which combines unfavorably with the synthetic hormones of the pill. It would be best to wait until more is known about the safety of these chemicals before testing them on your own body.

Drugs and Pregnancy

The mother-to-be should assiduously avoid any unnecessary medications, lest damage or deformity result to the fetus. Aspirin use increases the risk of brain damage, barbiturates have been known to cause bleeding in newborn babies, and antihistamines may produce breathing difficulties.

A generation ago diethylstibestrol (DES) was widely administered to pregnant women to prevent miscarriage. The hormone worked remarkably well and many children were born as a result, with no harmful side effects to the mothers. Strangely, when the female children reached adolescence, many of them developed a rare but deadly form of cervical cancer. Drug manufacturers had not conducted adequate laboratory tests, and in their haste had not even considered possible toxic effects on the unborn.

It wasn't until 1962 that medical science once and for all dispelled the myth that the placenta—the protective sac that surrounds the fetus in the uterus—effectively prevented dangerous substances from reaching the child. That year a cold chill of terror was felt around the world, when newspaper headlines revealed just how much that knowledge had cost. Irresponsible drug companies had promoted and sold a drug to relieve the discomforts of pregnancy, which actually caused a malformation of fetal arms and legs, making them resemble the flippers of seals. The name of the drug was thalidomide.

Thalidomide had been developed by a West German pharmaceutical company, Chemie Grunenthal, in the

1950s. Six years later, the William S. Merrell Company obtained the rights to manufacture and market the drug in the United States and Canada. Merrell geared up a high-pressure sales campaign, sent its New Drug Application to the F.D.A. for approval, and organized a group of doctors to use thalidomide in clinical investigation programs.

At the F.D.A., the application was given to Dr. Frances Kelsey, who told the company that its application was incomplete and did not adequately demonstrate safety. The company tried to apply pressure, but Dr. Kelsey was unrelenting. She had made a brilliant deduction that the company had overlooked: thalidomide induced sleep in human beings but not in test animals. Therefore, although the drug might be safe for animals, it was not necessarily safe for human beings. She asked Merrell to submit more information on the pharmacological properties of the drug and also human case histories.

Dr. Kelsey asked her husband, who was a physician-pharmacologist, to analyze the data the company sent her. He found that it was an "interesting collection of meaningless pseudoscientific jargon apparently intended to impress chemically unsophisticated readers."

Meanwhile, thalidomide was being widely sold in Canada, in spite of reports coming in from Europe that the drug was responsible for fetal malformations. Dr. Kelsey was convinced that the drug should never reach the American consumer. When *Time* magazine revealed in its February 23, 1962, issue that thousands of infants had been born in West Germany and England without arms or legs, or with only flipper-like appendages, Dr. Kelsey emerged as a heroine. Her courage and intelligence had saved many American women from living a tragic nightmare. In August of that year, President John F. Kennedy rewarded her with the Distinguished Federal Civilian Service Award "for her high ability and steadfast confidence in her professional decision."

Human Guinea Pigs

To our continued regret, effective "watchdogs" like Dr. Kelsey are few and far between in the drug industry and the F.D.A. When a new drug is put on the market, the consumer often is the guinea pig who must test its safety.

One tragic example is the drug that was known as MER/29, concocted by the before-mentioned William S. Merrell Company to lower blood cholesterol. These fatty residues are deposited on artery walls, and many statistics indicate that persons with higher than normal levels of cholesterol will develop hardening of the arteries, high blood pressure, strokes, and heart diseases. The drug went on sale in 1960, after being approved by the F.D.A., and Merrell launched a $1 million advertising campaign aimed at doctors.

Within months it was apparent that MER/29 did more harm than good. Some patients developed icthyosis (fishlike skin), a painful itchy malady, while others lost their hair, or developed cataracts. Ceasing to take the drug ended the skin disease and began a regrowth of hair, but cataract blindness was permanent. Information submitted by Merrell was subsequently discovered to have been falsified, and three executives were fined and sentenced to six months' probation. In civil suits, Merrell paid between $45 and $55 million to those harmed by MER/29.

The list goes on. Chymopapain, an injected drug that purported to repair slipped discs without surgery, was finally rejected by the F.D.A. after it had been used on 14,302 patients, several of whom died as a result. Serc, invented by Unimed, Inc., was claimed to be the first effective cure of Ménière's Syndrome, a progressively worsening condition of the inner ear that causes dizziness, ringing in the ears, and hearing loss. The F.D.A. seemed about to approve it when the Consumer's Union brought suit.[8] Serc was found to actually aggravate the symptoms of Ménière's Syndrome, and plans for its manufacture were dropped.

How do these drugs get through to the public, you might ask. First, physicians are overwhelmingly busy and cannot keep up with the flood of medical literature. This leaves them unaware of the information that might alert them to the dangers of a drug they are prescribing. Also, while most doctors know enough not to believe salespeople, too many of them believe the drug company's literature. And many place too much faith in the drug manufacturers' scientists, believing that the ethics of science preclude dishonesty on the part of the pharmaceutical researcher.

Ralph Adam Fine, a Department of Justice attorney and author of an excellent book detailing the MER/29 and thalidomide tragedies, sets forth two conditions that could go a long way toward diminishing the abuses of the drug industry. They are:

1. All new drugs, regardless of whether they are to be sold by prescription or over the counter, should be thoroughly tested before they are released to the public. Such testing must be conducted under the direct supervision either of the government, or of an independent group that has no financial interest in the test results.

2. All of the material pertaining to a drug company's experiences with a particular drug (except, of course, actual bona fide trade secrets) must be accessible to any interested individual. A physician cannot be expected to weigh the possible risks and benefits of a specific treatment without all the relevant information. "Puffing,"

which is a business euphemism for fraudulent advertising, must under no circumstance be allowed.

Notes

1. *The New York Times*, January 28, 1976, p. 1.
2. *Competitive Problems in the Drug Industry*, Hearings Before the Subcommittee on Monopoly, Senate Select Committee on Small Business, 90th through 93rd Congresses (1967-1974), parts 1-25.
3. Goodman, Louis S., and Gilman, A., editors, *The Pharmacological Basis of Therapeutics*, New York and London, 1965.
4. *The New York Times Encyclopedic Almanac*, 1975.
5. The Boston Collaborative Drug Surveillance Program, Boston University Medical Center, 1975.
6. Burack, Richard, with Fox, Fred J., *The New Handbook of Prescription Drugs*, Ballantine Books, New York, 1975.
7. The Planned Parenthood Federation of America, contraceptives study, 1975.
8. Consumer Reports, *Serc: A Dizzying Story of Vertigo in the F.D.A.*, March, 1973.

Fertilizers—A Crime
Against the Soil

All earthly life depends ultimately on plants—for plants are the only living organisms able to create life from inorganic matter. This is accomplished through the fascinating process of photosynthesis, by which plants utilize the sun's power in order to convert water, minerals from the earth, and carbon dioxide from the air into carbohydrates, proteins, and oils—the basic sources of energy of all living things.

Science has not been able to reproduce this deceptively simple chemical conversion in the laboratory. It has only gone so far as to develop various means of providing the plants with nourishment, so that they may draw an even greater supply of food from the soil.

Ordinarily we think of soil as "dirt," something we walk on that is muddy in winter and dusty in summer. But soil is where nutrition begins. Soil contains the raw materials that yield shimmering fields of wheat; firm, golden stalks of corn; and juicy, plump strawberries. Decaying rock particles and rock dust form the bulk of topsoil, the surface strata that is the growing medium of most food plants, the remainder of which is mostly a mixture of decaying vegetable and animal wastes known as humus.

Inside this rich topsoil is a complicated balance of living organisms that help the plant assimilate minerals and chemical compounds from rock particles. Funguses,

146

bacteria, earthworms, and insects are among the many forms of life that feed on humus. These minute animals slowly decompose plants left from the previous growing season, as well as animal carcasses and manure; they also serve to aerate the soil so that gases can be exchanged and water absorbed. As a result, sulfuric and carbonic acid are generated, which furthers the decay of rocks and releases their mineral contents, thus enriching the soil.[1]

In the wild, nature maintains a constant ecological balance. Plants that have created life from the soil return to it in death, as do animals that fed on plants. Absorbed and processed by the soil, the dead are recycled into the living. In nature there is no waste, no pollution.

Unfortunately, this perfect state no longer exists in the human food chain. For most of the 12,000 years since plants were first domesticated, farmers simply supplemented nature with organic fertilizers, and rotated crops or let fields lie fallow so that the soil's nutrients would not be depleted. The earth was not made to produce more than it was constitutionally able to bear.

Then, early in the nineteenth century, a renowned German chemist named Justus von Liebig discovered that plants could be artificially fertilized with chemicals. To determine the chemical elements needed by vegetation, von Liebig conducted a series of brilliant experiments through which he discovered the chemical substances used by plants. He burned numerous species of plants, analyzed the substances found in the ashes, and determined that soil was merely a mixture of these substances. If humans were to provide these chemical substances, he believed, plants would obtain all the nutrients they needed. As scientifically sound as this conclusion may appear, it failed to take into account that soil is more than its mineral content.

Von Liebig all but ignored the organic, living components of soil that are contained in humus. Being a laboratory chemist, he failed to understand that the myriad network of underground life—from moles, mice, and shrews to earthworms and microorganisms—is an indispensable, life-generating part of soil. To von Liebig's way of thinking, all that were needed were artificially produced nitrogen, phosphorus, and potash, three basic requirements of plants in natural form.

Death of the Living Soil

By the time von Liebig's artificial fertilizers became generally available, farmers in the United States had already robbed the land of one-fourth of its topsoil as a result of poor soil management.[2] The seriousness of this loss becomes readily apparent when we consider that it takes nature 500 to 1,000 years to replace a single inch of topsoil. Most of the early settlers and pioneers did not know how to conserve soil, and they did not bother to learn. After all, the land was free or very cheap, and there seemed to be a never-ending abundance of it. "Get what crops you can out of the land, and when it's burned out and can produce no more, move on," was their credo.

The wages of this random rape of the land were paid with a vengeance during the mid-nineteen thirties. Great dust storms boiled up over much of America's farmlands, blowing away clouds of black topsoil from recently plowed fields. The prairies had been overgrazed, trees which had once broken fierce winds and held moisture in the land had been cut down years before, and the earth was dried out from overcultivation. Thousands of impoverished farmers were forced to leave their wasted farms and migrate to the still fertile earth of California and the Pacific Northwest.

Today, most farmers, aware of the damage done by their ancestors, successfully combat destruction of the soil by wind and water erosion. At the same time, they have found a new way to destroy the land—by forcing it to produce more than it should with chemical fertilizers. Huge industrial farms, aptly dubbed "agribusinesses," have largely taken over the land of the small, conventional farmer who lived close to nature and consumed the crops produced. Today, the quantity of production is more important than quality, and most of America's farmlands have been polluted with artificial chemicals for the sake of profits.

Ecological balance no longer exists on most farms. Today's farmer tends to overplant a few limited crops, thereby depleting the soil of certain essential trace elements. In the past, a farm was a self-contained environment. Today, the produce farmer buys meat from the butcher shop and milk from a store or dairy farm, instead of keeping cattle, chickens, and pigs. Because agriculture has become so compartmentalized, the farmer has sacrificed a readily available source of natural fertilizer—animal wastes. This is an unfortunate loss, for soil dressed with manure produces crops that are more nourishing and tastier than those grown in chemically fertilized soil.

Chemical manufacturers insist on perpetuating the myth that there is not enough organic fertilizer to go around, but the facts do not bear this out. In fact, animal waste in the United States amounts to 2 billion tons annually, which is equivalent to the waste produced by half the world's population.[3] In other countries, manure is distributed to farms, an all-but-impossible task in the United States. Cows and pigs are concentrated in single feedlots that contain from 10,000 to 50,000 animals, and up to 250,000 chickens, and therein lies the problem. It would be prohibitively expensive to collect and transport all this natural fertilizer to fields where it is needed, thousands of miles away.

Bags of chemicals, therefore, become cheaper, cleaner and easier to transport. So, instead of contributing to the food chain by a natural recycling process, animal waste is disposed of as sewage to pollute the nation's water systems. In less than a century, humans have upset the balance of nature by robbing the soil of nutrients that are never returned to it. Even our waste is wasted.

The widespread application of artificial nitrogen, phosphorus, and potash (known to farmers as NPK) brings about changes in the composition of soil which destroy or seriously disturb organisms that benefit it. The presence of these organisms serves as a barometer of soil fertility. If they cannot survive, it is a sign that the soil will not bear crops worth eating. The work of earthworms and micro-organisms are essential, but they are destroyed by these chemicals. Super-phosphate fertilizers tend to create acid conditions in which they cannot survive. In Australia, nine-foot-long earthworms originally present in vast numbers were completely exterminated by this type of fertilizer.

The destruction of living things in soil occurs because the ingredients in artificial fertilizers are so readily water-soluble. In nature, easily soluble fertilizing elements rarely occur. For example, humus harbors plant nutrients that dissolve in water very slowly, feeding plants at a rate that precludes the possibility of poisoning them and their living benefactors in the soil.

Proper fertilization also involves more than the application of three concentrated chemicals to the roots of plants. More than a dozen minerals and trace elements are needed as well. Although these account for only one percent of a plant's needs, minerals and trace elements are extremely important nutritional factors. Many human diseases result from diets deficient in these factors, which are often not obtained from foods grown in chemically treated ground.

Chemical fertilizer manufacturers were quick to jump on the bandwagon when it was discovered that these elements were lacking in synthetic plant foods. They quickly mixed in a few, calling them such things as "power boosters," which didn't help the soil at all. All of these concoctions were totally imbalanced, for they did not stimulate a balance in the proportions that exist in nature. Consequently, the carbohydrate-protein ratio of many crops began to change for the worse, and vitamin content declined.

The Rotten Red Tomato

The sad state of the American tomato is a case in point. Once fragrant, flame-red orbs bursting with juice, tomatoes in recent years have become woolly, tasteless globules that can practically be bounced off the wall without being bruised. Fertilizers and hybrid strains combine to produce tomatoes that have superior handling and keeping qualities. But what about the loss of vitamin C and flavor?

No longer thinking in terms of patches and pounds, farmers were faced with new problems when production covered acres and amounted to tons. Harvesting machines would damage normal, tasty tomatoes, so a pulpy, thick-skinned hybrid that could withstand rough handling was created. Since agribusinesses have created a demand for fresh tomatoes the year around, the growing season has been unnaturally extended. Grown during the winter in southern and western states, tomatoes can no longer be left to ripen on the vine if they are to survive being shipped thousands of miles to the north. As soon as NPK forces them into existence, tomatoes are picked green and ripened artificially. During the long voyage in refrigerated trucks and trains, tomatoes are kept in temperature- and humidity-controlled environments that effectively stop their growth. Just before they are sent to your local market, tomatoes are sprayed with ethylene gas, which turns them red. The consumer is forced to purchase a nutritionally worthless, unripe, cosmetically treated product—or to do without tomatoes.

Excessive use of artificial fertilizers lessens the keeping qualities of many other food plants, making it necessary to pick them before they have absorbed whatever nutrients are left in the soil and ripen naturally. Industrial farmers fondly point to the beautiful, uniform appearance of their produce as proof of the benefits of NPK. But consumers are forced to eat celery that is as pithy as it is pretty, melon-sized and mealy cucumbers, and strawberries big as apples but with less flavor than the cardboard containers they come in.

The health of a plant is a complex matter that is not always reflected in its appearance. Crops regularly doped with chemicals never attain the optimum food value of their organic counterparts. The trace minerals mentioned earlier cannot be effectively absorbed, even when present in the soil. In artificially fertilized plants, the beneficial effects of humus are thwarted, if not destroyed. It is the finely dissolved particles of humus that transfer most of the minerals from the soil to root hairs. Being negatively charged, humus particles attract positively charged minerals, such as potassium, sodium, calcium, manganese, magnesium, boron, aluminum, iron, copper, and other metals. When nitrogen is poured into the soil year after year, both humus and root hairs become coated with it, and the transfer of minerals can no longer take place.[4]

Too much potash decreases synthesis of ascorbic acid (vitamin C), carotene (vitamin A), chlorophyll, and amino acids. Too much phosphorus produces a zinc deficiency. Livestock and poultry are fattened on chemically produced grain and pass these deficiencies on to us when we eat their meat. Humans, the last link in the food chain, inevitably suffer the consequences of this tampering with nature. Many medical researchers believe that the comparatively recent upsurge in degenerative diseases is directly related to the inferior quality foods produced by modern farming methods.

The Green Revolution

In the 1940s, the widespread use of chemical fertilizers and pesticides (which are dealt with in the following chapter) prompted the agricultural establishment to herald the arrival of a "green revolution." Super-hardy crops impervious to insect pests could now be grown in unending abundance, it was said, and the world's food shortages would soon be met. In the 27 years between 1946 and 1973, the use of nitrogen alone increased by over 550 percent. The total American use of NPK was nearly 45 million tons by 1973.[5]

Meanwhile, the protein content of farm crops began a steady downward slide that continues to this day. The promise of abundance was fulfilled, it is true, but at a heavy cost. In Kansas, for example, wheat yields per acre shot up dramatically when artificial fertilizers were introduced, but the protein yield declined in an equally dramatic curve. In 1940, Kansas wheat contained as much as 17 percent protein. In 1951, only eleven years later, the amount fell to 14 percent, the average yield being about 12 percent.[6] Starchy, cheap carbohydrates took the place of this life-giving foodstuff.

At a time when the world's hunger problems are particularly pressing, the "green revolution" has tried to meet the challenge with quantity, not quality. Chemical fertilizers weaken the proteins that remain by upsetting the delicate balance of amino acids within protein molecules. Their body-building, tissue-renewing qualities are seriously jeopardized. When a single amino acid is missing, as is often the case, the other nine refuse to do their job. If non-essential amino acids are not present, even though the others are, the essential ones may do only half their work. The body tries to compensate for these faulty foods by craving and eating more of them in order to meet its physical requirements. The eating of greater and greater quantities of protein foods which can be only partially utilized at best serves only to waste protein, which is not only in short supply in terms of the world's needs, but is the most costly item in the diet, as well.

Dr. William Albrecht, an internationally renowned agronomist, effectively sums up what is wrong with American agriculture when he states: "Man has become aware of increased need for health preservation, interpreted as a technical need for more hospitals, drugs, and doctors, when it may simply be a matter of failing to recognize the basic truth in the old adage which reminded us that to be well fed is to be healthy. Unfortunately, we have not seen the changes man has wrought in his soil community in terms of food quality for health, as economics and technologies have emphasized its quantity. Man is exploiting the earth that feeds him much as a parasite multiplies until it kills its host. Slowly the reserves of the soil are being exhausted."[7]

Deaf to such warnings, the chemical industry continues

to reiterate its blind faith in the ability of human technology to fabricate the solutions to our needs. It points to the fact that nylon and other synthetic fibers have largely replaced cotton and wool, that cars roll better on imitation rubber, that detergents wash better than soap, and that plastic dishes don't break the way china and earthenware do. One gets the impression that natural products will soon go the way of the horse and buggy.

What is seldom considered is the real cost of the manufacture of chemical substitutes for what the earth can produce more efficiently. For instance, both cotton and nylon consist of long chains of small units of molecules linked together (monomers). The cotton plant takes the energy it needs to produce fiber from the sun and draws raw material from the soil. It costs nothing and creates no pollution. Nylon, on the other hand, is made from petroleum—a fossil fuel that is stored plant energy of a millenia ago. To bind the molecules into the required monomers, petroleum or coal must be burned to supply energy to operate factory machinery. Thus, great amounts of non-renewable energy sources are lost forever. The factory produces air pollution as a byproduct of manufacture, and nylon and plastic gadgets, utensils, plates, and cups litter the landscape forever. They are new to the life cycle and no micro-organisms exist that can degrade and recycle them back to the soil.

Mention should be made here of the recent appearance of biodegradable products on the supermarket shelves—most notably, various laundry and cleaning products. Such products are able to be broken down into substances that can be recycled by the soil. This is one of the few ways in which today's consumer is at least given a choice between helping to preserve or destroy the earth's ecology.

Although the raw materials of chemical fertilizers are as

abundant as the world's stones and mountains, the manufacturing process consumes an immense amount of fossil fuel energy. Add to that the cost of transportation to farms and mechanical dispersal, and you have a truer picture of the tremendous waste that results from substituting the artificial for the real. In other words, the so-called "green revolution" consumes as much or more energy than it produces.

Streams of Disaster

Instead of returning our waste products to the land, where nature uses them as food, we simply get rid of it all as garbage. Untreated sewage eventually finds its way into America's streams, river, and lakes, along with disastrous amounts of NPK leached from the earth by irrigation and rain. Both provide nourishment for water plants and cause them to grow in abnormal numbers and sizes. This uses up the oxygen dissolved in the water, which in turn does two harmful things: it kills fish and other water animals that depend upon oxygen for life; and it takes away the self-purifying ability of the water. Dissolved oxygen acts on small amounts of pollutants such as sewage and changes them to pure, harmless substances. Even industrial wastes can be rendered harmless in small amounts by oxygen dissolved in ecologically balanced waterways. But massive doses of fertilizers have overwhelmed nature's defenses, and many bodies of water have suffered the fate of Lake Erie, which was, until recently, foul and practically lifeless.

According to the august Institute of Ecology, "It is a gigantic one-way flow of elements from the earth and the air into the sea. The scale of the operation is far greater than anything previously known on the face of the earth. And this human phenomenon is in stark contrast with the natural communities of plants and animals which have been living in balance with their surroundings for thousands of years."[8]

This one-way flow of essential elements from the earth to city to water can be stopped if a concentrated effort is made. Strong measures will have to be taken if we are to feed back the nutrients we now rob from nature. Organic fertilizers will have to be substituted for artificial ones; waste products will have to be processed and recycled if the closed system we live in is to survive.

The city of Chicago is a pioneer in reclamation of land and water—hopefully a harbinger of future progress in the United States. Working with the U.S. Army Corps of Engineers, Chicago's town planners recently set in operation a massive plan to clean up the badly polluted lower end of Lake Michigan. Tons of sludge obtained from sewage effluent and processed industrial wastes have already been spread over land ruined by strip mining and, within a few years, the land has nearly restored itself. By means of an innovative technique humans were able to speed up nature's slow process of regeneration.

So far the major portion of the plan remains to be implemented. Estimated at a cost in excess of $7 billion, the plan requires the building of huge rural lagoons into which sewage and waste would be pumped. Aerated, the lagoons would contain bacteria that break down wastes into organic fertilizer. Several times a year the fertilizer would be collected from the bottom of the lagoons (where it settles) and sold to farmers at a low price. The fresh, cleaned water would be pumped to farms for irrigation, where it would be cleaned again by the living soil before flowing back into waterways. Organic nutrients would remain in the earth, and no pollution would result. If numerous communities were willing to adopt such a plan this could be an important first step in restoring the earth's ecological cycle. Perhaps our produce, fish, and water could one day again be safe to eat and drink.

A great deal of controversy surrounds the plan, however, and it may be years before it is put into effect. People don't want lagoons to be built in their vicinity, for they falsely equate lagoons with swamps. Through lack of knowledge, they think of such unpleasant things as mosquitoes, malaria, and snakes. Some farmers mistakenly believe that their crops would be fertilized and watered with "Chicago's filth." If an intelligent explanation of the causes of environmental and body pollution were to be promoted and advertised as widely as the products of industrial manufacturers, such misunderstanding would not exist. But there are no dollar profits to be gained from informing the public. Obviously, health and the elimination of the root causes of many diseases are not even minor considerations.

Bringing Down Baby

The subtle and gradual poisoning of the land has been completely overlooked by most farmers in their rapid and unquestioning acceptance of chemical farming. It took a few years before the detrimental effects became evident and by that time the fertilizer industry had become so large there was no stopping its crushing wheel of "progress."

The prime component of chemical fertilizers is nitrogen (nitrates) produced in the laboratory. Runoff water containing nitrates often seeps into farm ponds and wells, rendering them unfit for human and animal use. Cattle drinking nitrate-contaminated water lose weight; they are no longer able to completely utilize their feed. Cows show the symptoms of nitrate poisoning by giving less milk, and what they do produce is of inferior quality. If not treated at once, animals and humans soon die.

Nitrogenous fertilizers have their most immediate and drastic effect on babies. The source can be either polluted water, or vegetables that have absorbed too much of the fertilizer. Public health officials are alarmed at the increasing occurence of a disease, methemoglobinemia, the cause of which is directly linked to nitrates. In 1945, it was discovered that certain bacteria in the stomach are able to convert nitrogen compounds into poisonous nitrites, similar but deadly. When nitrites enter the bloodstream, they react with hemoglobin (the red pigment in the blood) to form methemoglobin. Since hemoglobin carries oxygen to tissues via the blood and methemoglobin does not, the victim may turn blue and, in some cases, suffocate and die. Infants are particularly susceptible to this form of poisoning, for although their stomachs contain less acid than do those of adults, their intestinal flora contain certain types of bacteria that facilitate the transformation of nitrate to nitrite.

In 1945, only two cases of methemoglobinemia were reported in the United States. In 1950, several years after the introduction of chemical fertilization, 139 cases had been identified in Minnesota alone: 14 of these cases were fatal.[9] In 1962, the Public Health Service recommended that when the amount of nitrates in drinking water reaches 45 parts per million, public warnings should be issued and parents urged to give their children bottled spring water instead of tap water—an expensive and inconvenient solution for most people. Unfortunately, there is no simple test an individual can use to determine whether this danger level has been reached.

Further scientific studies continue to show that food as well as water can be dangerously contaminated by nitrates. In fact, many baby foods contain lethal amounts of these chemicals. A 1971 study at the Missouri Agricultural Experiment Station came up with the evidence that several brands of canned baby food contained as much as 40 milligrams per two-ounce jar. This amount is well in excess of the 12 milligrams of nitrogen as nitrate recommended as a maximum daily consumption limit for infants by the Public Health Service.

Nitrates saturated in soil tend to accumulate in the leaves and stems of certain plants, especially spinach, beets, and carrots. Canners of these vegetables are plagued by the problem of internal corrosion of the cans caused by an excess of nitrates. These days if the cartoon character Popeye were to eat a steady diet of canned spinach to make himself strong, he would probably turn blue instead. And the old parental admonition would have to be changed to: "Eat your spinach! It'll make you sick."

Still another harmful effect of the high level of nitrates in drinking water and produce is its ability to induce cancer-causing substances known as nitrosamines. Researchers have noted the high incidence of stomach cancer in Japan, Chile, and Iceland, where large quantities of fish are eaten. These fish contain high levels of nitrates because the waters are rich in chemical fertilizers leached from the land. (Other factors which may contribute to the incidence of stomach cancer are the high content of polyunsaturated fat in fish, and traditional habits of eating rapidly.) For Americans, however, nitrosamines present more of a problem in their indiscriminate use by food processors as a preservative for such foods as luncheon meats, salami, hot dogs, ground beef, ham, bologna, and frankfurters. They are discussed more thoroughly in the chapter on food additivies.

The Stop-Growth Trick

After years of forcing food plants into producing nutritionally inferior abundance, agribusinesses have learned to play a new trick on nature. Just before harvesting, a chemical is sprayed on the leaves of crops that sends them into a state of suspended animation. Absorbed by the foliage, the chemical travels down the stem to the root and stops further cell division—meaning growth. By doing so, the farmers can pick crops at their leisure, and no longer have to wait until harvest time, when vegetables have reached their peak of flavor and nutrition.

There are more than 40 growth-regulating chemicals approved for use by the United States Department of Agriculture, and you can be sure that more are on the way. Maleic hydrazide is used to delay potatoes and onions from sprouting until they get to market, perhaps months later. The growth of sprouts ("eyes") on potatoes and a blossoming stem on onions is part of their natural aging process, which formerly enabled consumers to tell whether they were fresh. Treated this way, all supermarket produce looks fresher than it indeed is.

Maleic hydrazide has been known to cause damage to the liver and nervous system of experimental animals, as well as chromosome damage that results in deformed offspring. Not content with merely deceiving us, growing inferior crops, and polluting the environment, industrial farmers rub salt in our wounds by slipping yet another undesired chemical into so-called "fresh" food. Like most chemicals approved for use by our government, sprout and growth inhibitors have not been adequately tested for safety. Just the fact that these chemicals are able to *stop or delay* the life-processes of a living organism should be cause enough for alarm. The Department of Agriculture should be conducting experiments to answer the question posed by the chemicals' function: Couldn't growth inhibitors also arrest human growth? Until that question can be answered without any doubt, these chemicals should be banned entirely.

Still other chemicals are poured over plants to increase their size. Americans are particularly proud of the size of their produce, and agricultural fairs give prizes to the biggest cabbages, melons, and tomatoes, even though they are not the best. Inside, treated vegetables and fruits tend to be mealy and tasteless, having used up the nutrients they were designed to produce in their own struggle for outrageous growth.

Grapes, for example, are treated with a potentially harmful hormone, gibberellin, which turns them into freak growths. Large and solid, with elongated bodies that adhere tightly to the stems, such grapes lack flavor and juice, which is why many California vintners refuse to use them. Having been bred and conditioned to harvest easily (grapes that fall from the vine are hard to retrieve), they are no longer desirable for human consumption.

Fruit Cosmetics

Citrus fruits are often colored with dangerous coal-tar dyes—like the infamous cancer-causing Red dye No. 2, which was recently banned—to make them look more palatable in the market place. Showing a natural green on the rind of an orange is synonymous with loss of sales according to the citrus fruit industry, and the offending spots must be rouged over. Again, the quality inside is a minor consideration, and the consumer who grates orange peel into food must ingest a potentially harmful chemical.

Pears, apples, plums, and other fruits are coated with wax and mineral oil to make them look more attractive, and to improve their keeping qualities. The F.D.A.

approved the direct application of these substances on fruit and vegetables in 1964, after years of permitting only their packaging to be coated. That the wax used was the same as that used to polish floors seemed to cause the guardians of the nation's health little concern. Medical researchers at Johns Hopkins Hospital estimate that fruit eaters ingest about 50 grams of mineral oil yearly, which they believe may account for many unexplained tissue injuries. Germany banned such coatings in 1938.

Much of this produce is further sprayed with a toxic group of chemicals known as phenols to preserve them during the long journey to the market place. Even in doses as small as one and a half grams, phenols are so lethal that they can induce vomiting, circulatory collapse, convulsions, and decay of the mouth and intestinal tract if swallowed. When phenols were first approved for use in the 1950s, the F.D.A. stipulated that a sign would have to be placed next to the container of the treated produce. The wording, "To Maintain Freshness," was another example of how the food industry will subtly twist the truth and convert a warning into an enticing invitation to buy. Under pressure brought to bear by agribusinesses, the F.D.A. relaxed its ruling and the warning card was discarded. American consumers, it seemed, were again being sold out by the protective agency supported by their taxes. Only the Germans and Italians are given that consideration by their governments, which have passed laws that require phenol-treated American citrus fruit to be stamped with the following words "With Diphenyl. Peel Unsuitable for Consumption."

Can a Catastrophe Be Averted?

The widespread use of chemicals to grow and treat the raw abundance of nature is truly disheartening. As long as we rely on artificial fertilization, it is only a matter of time until all our soil is made useless for growing crops. Ecologists are worried about the arrival of a time when there is no more land to cultivate. It has been estimated that the total destruction of fertile soil and the accompanying disappearance of all plant life on earth would mean the extinction of all animal life within one year. This frightening fact is perhaps the best expression of how dependent humans and other animals are on the proper and natural use of soil.

There is no way to circumvent the life cycles of the soil for very long. They are intrinsic, essential, and far-reaching. We can disrupt them, but if we do, we cannot prevent the ensuing devastation that will be an inevitable result. Only through organic agriculture—working with natural materials that are the core of soil structure—can we cooperate with nature in an intelligent and fruitful manner.

It is time for us to realize that we are only one small part of the food chain. We can never control it, but we can destroy it.

Notes

1. Balfour, C.B., *The Living Soil*, Devin-Adair, New York, 1952.
2. Borgstrom, Georg, "Food and Ecology," *Ecosphere*, The Magazine of the International Ecology University, 2 (No. 1): p. 6, 1971.
3. *Environmental Science and Technology*, 4 (no. 12): p. 1098, 1970.
4. Gerras, Charles, and others (editors), *Organic Gardening*, Bantam Books, New York, 1972.
5. *The Statistical Abstract of the United States*. Grosset & Dunlap, 1975.
6. Thomas, Jr., William L. (editor), *Man's Role in Changing*

the Face of the Earth, University of Chicago Press, 1956.
7. *Ibid.*
8. *Man in the Living Environment,* The Institute of Ecology Report on Global Ecological Problems, 1971.
9. *Journal of the American Water Works Association*, June, 1950.

People or Pesticides—
A Life and Death Battle

Every day countless numbers of us suffer the effects of pesticide poisoning without realizing it. We attribute our headache, fatigue, nervousness, aching bones, and fever to a cold or flu—and the true cause never occurs to us. Doctors rarely recognize these symptoms for what they really are, as there is so little available research to guide them. And the possible link between pesticides and birth defects and cancer has been virtually ignored by the medical profession. As Dr. Samuel S. Epstein—a cancer expert at the Harvard Medical School—has pointed out, pesticides "could affect and catastrophically so, as many as 1/10,000 of the population and yet probably escape detection by conventional procedures."[1]

Human beings are being used as guinea pigs to test the thousands of commercial poisons that are liberally sprayed on crops to control bugs, funguses, rats, weeds, and other forms of life that commercial farmers consider to be pests. These chemicals are dangerous to humans for they pollute the air we breathe, the water we drink, and the food we eat. The human body has had no previous experience with these synthetic chemicals, and there is no natural machinery in the body to break them down, much less eliminate them. The National Academy of Sciences reports that an average American consumes approximately 40 milligrams of pesticides each year in food alone, and carries about one-tenth of a gram permanently in his or her body fat.[2]

The need for pesticides is a direct result of human ecological ignorance. In a race to plant and prosper, humans have destroyed vast forests and plowed under lush grasslands, replacing them with single-species crops such as wheat, corn, and beans. Most living things were banished from the soil, or killed by artificial fertilizers, and the self-governing harmony of nature was completely lost.

Entire crops were often devastated by grasshoppers, corn borers, and boll weevils—pestilences the human race had brought upon itself by its utter disregard for the balance of nature. In the wild, the myriad species of insects had been kept from proliferating because they preyed on each other. Insects were also a basic food supply for field animals like frogs, lizards, snakes, and birds. In addition, many bugs were actually beneficial to plants and, like bees, served the ecological function of pollination, which produced seeds and fruits. Others fed on decaying organic matter and recycled it back into soil nutrients.

Granted the rare opportunity of an unlimited food supply that stretched over miles of cropland, with most of their natural enemies removed or destroyed, several species of insects increased at an explosive rate abnormal in a natural environment. The expansion of international trade after World War II made matters worse when new crops were introduced to countries where they had never before been grown, for new insects came along as uninvited guests. Many of these newcomers had no natural predators, which had helped control them elsewhere, and they multiplied even faster than their indigenous cousins. Furthermore, the widespread use of artificial fertilizers produced many genetically weak plant strains that were especially vulnerable to insects and funguses.

Rather than correct the problem by restoring to nature some of the things they had taken away, farmers resorted to the further use of chemicals. Over the last 33 years, billions of tons of synthetic pesticide compounds have been spewed into the environment. Blind to all considerations except yield per acre—meaning profits—chemical manufacturers conducted practically no research as to the effect of these sprays on other forms of life, on the ecology, or on the soil. As a result, we have paid the price of our good health for this criminal negligence.

DDT

Touted as a "miracle" pesticide by Swiss manufacturers who introduced the synthetic compound to the world in 1942, DDT is one of the most potent poisons known to humans. An obscure German chemist created DDT (dichloro-diphenyl-trichloroethane) in 1874 and it was considered useless until rediscovered by a team of Swiss scientists who were searching for an effective insecticide. Tried experimentally on a Swiss potato crop that was being ravaged by the Colorado beetle, DDT eliminated the bugs and all other predators as well. The poison was remarkably effective because it acted in three ways: on contact, in the stomach, and in the lungs.

During World War II, DDT was sprayed in powder form on American troops, their prisoners, and the civilian populations of the territories in which the troops fought, effectively preventing outbreaks of typhus and cholera, which had always occurred during previous wars. Malaria-carrying mosquitoes were nearly eradicated on South Pacific islands and public health officials predicted the day when malaria would no longer exist. For what seemed a great benefit to humankind, Dr. Paul Herman

Muller, the Swiss scientist who perfected DDT's use as an insecticide, was awarded the Nobel Prize in 1948.

When the war was over, DDT's proven effectiveness found a peacetime application and it was used by crop growers to rid their fields of insects. DDT destroyed insects by the billions, and it was sprayed into nearly every corner of the world during the next 27 years. A major contributing factor to its popularity was DDT's extremely low price, which dropped from $1.19 a pound in the early 1940's, to 17 cents a pound a few years later.[3] Few people were alarmed by this widespread contamination of the earth, for not even scientists were aware of DDT's long-term effects. Most praised DDT as a "completely safe," conveniently cheap, super-strong insecticide that would eventually rid the world of pestilence, and subsequently hunger as well.

Some researchers disputed this unrealistically optimistic view of the scientific community, which had been formed at least partially by pesticide industry propaganda, but their valuable—and prophetic—findings were virtually ignored at the time. In July of 1945, the *British Medical Journal* released the results of a study in which experimenters applied DDT directly to their skins. Most of them suffered symptoms of chronic tiredness, irritability, indeterminate body pains, and a "feeling of mental incompetence in tackling the simplest mental task." One researcher was so severely affected by DDT poisoning that he was bedridden for ten weeks.

Reports of DDT's toxic effects on humans and other animals surfaced occasionally in medical and scientific journals, but it was not until 1950 that a Congressional investigating committee revealed to the world that DDT was not the blessing it had first seemed. Testifying before the Delaney Subcommittee, which was attempting to ban food adulterants, Dr. Morton S. Biskind of the American Medical Association issued the following warning:

"The introduction for uncontrolled general use by the public of the insecticide 'DDT'...and the series of even more deadly substances that followed has no previous counterpart in history. Beyond question, no other substance known to man was ever developed so rapidly and spread so indiscriminately over so large a portion of the earth in so short a time. This is the more surprising as, at the time DDT was released for public use, a large amount of data was already available in the medical literature showing that this agent was extremely toxic for many different species of animals, that it was cumulatively stored in the body fat, and that it appeared in the milk."

Dr. Biskind went on to explain that a large number of cases of "virus infections" and "intestinal flu" are actually incidences of DDT poisoning from foods that contained larger-than-usual residues of DDT.

Under pressure from the Delaney Subcommittee, the F.D.A. began to test milk products for residues of DDT, which the agency claimed would not be allowed on the market. To its surprise, the F.D.A. discovered that 62 percent of all milk contained DDT; so did 75 percent of the butter and 50 percent of the cheese. Clearly it was hopeless to try to keep DDT and other pesticides out of milk at this late date, so they set tolerance levels for these poisons. The Government allowed 0.05 parts per million (ppm) for whole milk, and 1.25 ppm for milk fat—cream, butter, and cheese. These tolerance levels are still in effect, but they are rarely, if ever, adhered to. Today, it is impossible to drink milk distributed through a commercial source without polluting one's body with an unsafe amount of DDT.

The F.D.A. did prohibit the use of DDT in and around

dairy barns, where it had been sprayed directly on the animals for years, since it was discovered that even if DDT was applied to the barns while the cows were outside and the feeding troughs covered, the poison still showed up in the milk the next day. But the cattle feed was already contaminated with DDT, and air-borne sprays often wafted to the barns, so all the F.D.A. had done was to lower the amount of contamination, not eliminate it.

Contaminated milk is a serious problem for infants, who are especially susceptible to pesticide poisoning. Since milk is a basic food for babies, and since most babies have a large amount of body fat, DDT is stored in their tissues in much greater proportions than in adults. When a baby learns to walk, it uses up its reserve of fat, and the stored DDT is released into the bloodstream, where it has been detected in high concentrations.

Breast-feeding is no solution, for mother's milk also contains high levels of DDT which has been absorbed by the mother from food and the environment. Women, especially if they are lactating, possess a larger amount of body fat than men do, and consequently absorb more DDT. Random samplings have shown that human milk can contain as much as 116 ppm of DDT, which is 2,320 times the legal limit set for cow's milk.[4]

DDT enters the food chain through the soil, where it remains in an active state for up to 40 years. It is a very stable chemical, not easily decomposed or broken down. Tests of soil powdered with DDT retained 80 percent of the chemical seven years later. Since farmers have been spraying crop fields for as many as 26 years, each year's amount has accumulated, steadily saturating the soil with an enormous amount of poison. Thus, we and future generations will be eating DDT for years to come.

Plants absorb minute amounts of the chemical in

growth, and pass it along virtually unchanged to the
creatures who eat them; the plants also contain surface
residues that are ingested. To produce one pound of meat,
livestock must eat approximately 16 pounds of feed,[5]
which means that huge quantities of DDT are daily
absorbed by these animals and passed along to humans,
who are at the top of the food chain. The fat-marbled steak
that meat-eaters are so fond of is literally a *dangerously*
expensive way of obtaining a daily supply of protein.

The DDT stored in the body is not confined to a
particular area or organ where it cannot harm the rest of
the body. The cells accumulated as fat are constantly
undergoing metabolic changes, and they are involved in
the "inner-ecology" of all the body's functions. Under
conditions of physiological need or stress, fat is released
into the bloodstream, and so is DDT. A sudden weight loss
from illness or dieting can saturate the tissues with this
poison, and cause symptoms ranging from uncontrollable
tremors, to loss of appetite, a jaundiced coloring, and
fever. Not much is known about the long-term effects of
DDT poisoning because the chemical is relatively new to
the environment, but the evidence available is a frightening portent of what may be currently happening to many
of our bodies.

Testifying in a court case in 1958, Dr. Malcolm
Hargraves, a Mayo Clinic blood specialist, stated that he
was positive that DDT caused leukemia, Hodgkin's
Disease, anemia, and other blood disorders; Dr. Hargraves' educated opinion was based on autopsies and past
case histories. All of these diseases are related to liver and
spleen functions, and all have increased at a spiraling rate
since DDT came into general use in 1945. The incidence of
leukemia is now the highest in farm states that have
sprayed the most DDT.

As the detoxifying, filtering organ of the blood, the liver must constantly work overtime to cleanse the body of foreign, inorganic substances like pesticides, food additives, and air pollutants. Its main job of straining out the natural toxins that occur in many foods may be seriously interfered with because of this extra burden. DDT has been known to destroy vitamins and enzymes necessary to the proper functioning of the liver and spleen, seriously weakening these vital organs and making them susceptible to viral and bacterial infections. Both hepatitis (inflammation of the liver) and cirrhosis (scarring of the liver) have increased at an alarming rate over the past two decades, and not only among drug users and alcoholics—the traditional victims of these diseases.

The problem of DDT contamination was eventually brought to the attention of the United Nations, which conducted a two-year study under the auspices of its subsidiary, the World Health Organization (WHO). Monkeys were given oral doses of DDT at the rate of 0.2 milligrams per kilogram of body weight for periods of seven to nine months, and most developed hepatitis. The monkeys who did not were continued on the DDT diet and, after a year, they showed signs of liver enlargment and hyperglycemia (high blood sugar), a pre-diabetes condition. In its report, released in 1966, WHO speculated on DDT's possible application to human beings:

"The possibility that [the bodily accumulation of DDT] might have deleterious consequences later in life cannot be ruled out. In addition, it has not been demonstrated that metabolic changes in liver cells comparable to those observed in [laboratory animals] do not take place in man. At the present time, [the United Nations is] concerned about the storage of DDT which occurs in all species and about the cellular changes produced in the liver of rats by

DDT and by other compounds chemically related to it."

The U.S. was also concerned by the mass slaughter of birds that had taken place in the years since DDT was introduced. Naturalist Rachel Carson brought this tragedy to the attention of the world in her best-selling book, *Silent Spring*, published in 1962, when she described the grave-like stillness of verdant farmlands that were totally devoid of birds. Within the space of a single generation the bird populations of the United States and Europe have been decimated to the point where many species might soon be extinct.

The familiar, reassuring chirp that signalled the beginning of spring and the return of birds from their winter migration had disappeared along with the so-called "pests" destroyed by DDT. The natural enemies of insects that prey on our food plants, birds descend on crop fields to feast on the bugs they find there, whether dead or alive. From the 1940s on, robins, sparrows, and other birds have been observed plummeting from the sky, acutely poisoned by lethal doses of pesticides ingested from bugs, worms, and other sources of polluted food.

The birds who have survived this chemical assault face another equally disastrous problem: they are unable to reproduce. And when they do manage to lay eggs, the few chicks that do hatch often turn out to be deformed or weak mutants who are unable to survive.

By 1965, the peregrine falcon, the brown pelican, and the double-crested cormorant had all but disappeared from their natural habitat along the coast of California. The bald eagle, our national symbol, had stopped breeding in its favorite nesting grounds along the Great Lakes and the Atlantic Coast. According to a five year study conducted by the National Audubon Society, the problem was that their eggs were not hatching. Of 53 nests surveyed in Maine only four produced live bald eagle chicks, an astonishingly low birth rate unheard of before the widespread use of DDT.

Researchers discovered that the chemical is absorbed and concentrated in a bird's bloodstream in much greater quantities than in a human's, especially at egg-laying time. Containing a higher quantity of lipids (fatty substances) that facilitate the hardening of eggs, a bird's bloodstream is a perfect medium for fat-bonding DDT molecules. As the poisons accumulate, a bird's ability to produce hard shells seriously diminishes; many eggs are so thin and fragile that they break apart when the parent sits on them in the hatching process In 1969, Audubon Society ornithologists discovered the ultimate result of this process of deterioration: an egg with no shell at all, only a transparent membrane that covered the dead eagle embryo.

Birds that feed on fish and water creatures have suffered the same fate as land-feeders, since most of our waterways are badly polluted by pesticides. Flowing from irrigated, rain-washed land, DDT is ingested by all fish, who pass it along to birds and mammals—including humans—who eat them. The Study of Critical Environmental Problems (S.C.E.P.), convened by the Massachusetts Institute of Technology, estimated that more than one-fourth of the 63,400 metric tons of DDT that chemical manufacturers produced in 1968 has found its way into the oceans.

Like all other food chains, the ocean's food chain begins with the smallest forms of life. Therefore, by the time a large fish has reached maturity, it has consumed huge quantities of smaller ones; this means that it has also saturated its tissues with the DDT stored in the bodies of the smaller fish. Thus, it is wise to restrict one's fish intake to the smaller varieties, although this is no guarantee of purity. Clams, oysters, shellfish, and crabs have all been found to contain extremely high levels of pesticide contamination out of proportion to their size.

Other creatures suffer from ocean pollution as well. Concerned over the recent decline in the population of sea lions that live on the California coast, naturalist groups traced the cause to their sea-food diet. Being enormously fat (nature's way of ensuring this warm-blooded mammal's survival in cold oceans), the sea lions had absorbed huge amounts of DDT, which caused females to lose their mothering instincts. Cows were observed to bark at and push their pups into the ocean, rather than rounding them up and returning them to the herd.

DDT Substitutes—Deadlier Yet!

As soon as it became apparent that many species of insects were developing a resistance to DDT, chemists created more effective poisons. Most of these substitutes are very similar to DDT, and they belong to the same chemical family of chlorinated hydrocarbons. These include aldrin, BHC (also called lindane), chlordane, dieldrin, endrin, heptachlor, methoxychlor, and perthane. Their toxic effects are approximately the same as DDT, and accidental ingestion of the pure chemicals, or massive body contact, can cause nervous excitation, convulsions, coma, and possibly death. Each year numerous children and adults die when they eat directly contaminated food, or drink from bottles used to store chlorinated hydrocarbons. Excessive exposure to these innocuous looking powders from improperly stored, open containers causes additional fatalities among farm workers, who inhale the pesticides or absorb them through their skins.

According to Dr. Malcolm Hargraves of the Mayo

Clinic, lindane is the most potent member of this group, for it depresses the bone marrow and prevents the formation of red blood cells. In one case, a man suffering from uncontrollable tremors, extreme fatigue, and headaches, was diagnosed as having a rare blood disorder that would soon prove fatal. Mayo Clinic investigators discovered that an exterminator had recently sprayed the basement of the man's house with lindane to kill a termite infestation and that the chemical had gotten into his gas furnace. For several weeks he had been inhaling the fumes from heating ducts that distributed lindane throughout the house. It took a week to decontaminate the premises. The man, however, was bedridden for over two years while his body decontaminated itself.

Another of Dr. Hargraves' cases involved a young woman who was being treated for aplastic anemia, also a rare blood disease, which requires blood transfusions every few weeks for the rest of the patient's life. Suspecting pesticide poisoning as the causative agent, Dr. Hargraves traced the source to an attractive, small, "decorator-look" lindane vaporizer kept on the woman's piano where she practiced several hours a day. Bothersome mosquitoes and flies had been killed on contact, and the woman had been poisoned by the fumes.[6]

Home pesticides are no less lethal than those used in fields. That colorful can of bug killer kept in the utility closet, with its artificial floral fragrance, is as effective against a human being as it is against mosquitoes and flies. A spray or two can be fatal to a curious three-year-old, and constant exposure to bug-bomb mists and vapor-emitting "pest strips" guaranteed to "kill all day" can seriously interfere with a child's physical and mental development. Ordinary, old-fashioned flypaper, for all its unattractiveness, is the safest means of destroying flying insects in the house.

Chemical manufacturers call these poisons "broad-spectrum treatments," a euphemism for the fact that they kill every living thing they come into immediate contact with, and they try to minimize the harm done to wildlife and humans by using the polite term, "side effects." Even the word "pesticides" is designed to make us think favorably of these chemicals; the reality is that DDT and its ilk kill *all* insects, only some of whom are regarded as pests. Scientists more correctly describe pesticides as "biocides," meaning that they destroy the web of life by upsetting the balance that nature took millions of years to achieve.

Few of us realize that insecticide manufacturers are owned by giant corporations that also own huge, mechanized farms and food processing factories. Today's farmer is actually a financial conglomerate looking for fast, large profits from every arm of its vast, tentacle-like grip on our food supply. As long as artificial fertilizers and pesticides continue to produce three-way profits for them, it is unlikely that agribusinesses will voluntarily stop the mass pollution caused by chemical residues. Corporate giants sponsor pesticide research programs at many universities, which suggests scientific respectability but only provides paid "experts" to justify the use of insecticides. The federal and state governments support these same agricultural colleges, which means that Americn taxpayers are unknowingly contributing to the invention of new chemicals that will eventually be absorbed into their bodies. In California, where 20 percent of the nation's insecticides are used, five of the eight largest federal government farm subsidies for 1970 went to huge agricultural conglomerates.

"The governing principle for the development of chemical pesticides," writes environmentalist Ralph

Nader, "seems to be selling pesticides rather than controlling insects."[7] To counter growing accusations that they had shown a total disregard for public health, and to promote the lie that pesticides are benign substances to human beings, manufacturers formed the National Agricultural Chemicals Association. The hero of this public relations organization is Dr. Wayland J. Hayes, Jr., former chief of toxicology of the U.S. Public Health Service, who in the 1950s conducted a much quoted and reprinted study which told the industry what it wanted to hear.

Obtaining permission from the Bureau of Prisons to test the effects of DDT on a group of 51 volunteer inmates, Dr. Hayes fed the men up to 35 milligrams of the pesticide each day for periods of up to 18 months. The dosage was approximately 200 times the amount found in the average American's diet at the time, but somehow no adverse effects turned up. According to Dr. Hayes, the prisoners began to eliminate DDT from their systems after about a year, which indicated that "a large safety factor is associated with DDT as it now occurs in the general diet." Only two of the men complained of ill health, one of whom was afflicted with headaches, watering of the eyes, and bone pains, which Dr. Hayes dismissed as "obviously . . . of psychoneurotic origin."

By prevailing rigid scientific standards, however, Dr. Hayes' findings could in no way be considered conclusive regardless of what the chemical industry wanted the public to believe. First, only five prisoners completed the full 18 months of the study, most having dropped out within the first year. Second, only 35 of the 51 men returned to have their urine and fatty tissues examined for the final tests.[8] Despite its highly subjective findings from such a small sampling, Dr. Hayes' study was widely

publicized as the final word on the subject by the
N.A.C.A., and for many years effectively countered the
well-documented claims of noted conservationists.

When an approval is sought on new insecticides, the
powerful N.A.C.A. lobby supplies government regulatory
agencies with its own research. Like the F.D.A., the
United States Department of Agriculture (U.S.D.A.) too
often seems to favor the interests of the industry it is
supposed to regulate. This betrayal of the public trust
became apparent in the celebrated court case that a
citizen's group known as the Environmental Defense
Fund (E.D.F.) entered against the Department of
Agriculture on behalf of the American people.

Backed by groups such as the Sierra Club and the
Audubon Society, the E.D.F. brought suit to force the
U.S.D.A. to totally ban the manufacture and use of DDT
and some of its analogues. Under the terms of its charter,
the only way the U.S.D.A. could do this was to rescind the
registrations of pesticide manufacturers, which gave it a
convenient out; U.S.D.A. decisions on the safety of
agricultural chemicals, the agency maintained, could only
be challenged by registrants. In other words, only
pesticide manufacturers could ask to have their products
taken off the market, and the public could have no say in it.

On the grounds that it would be impossible to remove
pesticides from the American diet, the F.D.A. turned
down the Environmental Defense Fund request that a
zero tolerance level for DDT be placed on all food. By
setting "safe" tolerance levels, the E.D.F. argued, the
agency was in effect legalizing an additive that its charter
did not cover in the first place, since it was allowed only
the power to ban additives included in food processing.
The renowned anthropologist Margaret Mead summed up
the dilemma by comparing it to *Alice in Wonderland*. By

"deliberately poisoning our food, then policing the result,"
she stated, the government is acting like Lewis Carroll's
White Knight, who thought of "a plan to dye one's
whiskers green, and always use so large a fan that they
could not be seen."

Unable to defend its industry oriented point of view
when faced with hard scientific evidence accumulated by
the environmentalists, the U.S.D.A. lost its case. Finding in
favor of the plaintiffs in May, 1970, Chief Justice David
Bazelon of the U.S. Court of Appeals for the District of
Columbia ruled that "the interest of the public in safety" is
as valid a reason for challenging the decisions of an
administrative agency as "the economic interest" of
pesticide manufacturers.

The Secretary of Agriculture was ordered to either ban
DDT within 30 days or bring new evidence to the court
that would justify its continued use. The agency cancelled
the use of DDT and several other chlorinated hydrocarbons on most fruits and vegetables, trees, livestock, and
buildings, but allowed its continued free use on citrus
fruits and cotton plants. The effect was the same as if
nothing had happened, however, for pesticide manufacturers simply appealed the decision in the courts and
continued to sell their poisons while they waited to come
to trial.

The ban was further delayed when President Nixon
brought into existence the much-needed Environmental
Protection Agency (E.P.A.), which took over the function
of pesticide control from the Department of Agriculture.
This meant that the E.D.F. had to re-file its suit because
the previous ruling did not apply to the new agency. In
January, 1971, the same court handed down an even
tougher decision that simultaneously cancelled the
industry's protest and ordered E.P.A. boss William

Ruckelshaus to call an immediate halt to *all* DDT spraying. Ruckelshaus accepted without contest, but his avowed get-tough, consumer-protection policy proved ineffective, for again a delay was granted and exceptions made.

Spraying of fields with DDT was finally prohibited two years later as of January 1, 1973, except for "public health purposes"—meaning in cases of epidemic disease—which may well turn out to be a legal loophole. The green peppers, sweet potatoes, and onions that you buy at the supermarket are still treated with DDT in storage on the excuse that no effective alternatives are available. For all the efforts made on its behalf, the public had won only a minor skirmish in its battle for unpolluted food. But at least they had fought back and shown their strength. The agricultural establishment and the government had been warned that it is the people who must be served, not the other way around.

In 1975, more than two dozen employees of a Hopewell, Virginia, chemical manufacturing plant brought suit against their employer and several government agencies, charging that their health had been seriously endangered by working with an untested pesticide called Kepone (pronounced Key-poone). Used in more than 40 household products for control of roaches and ants, and exported all over the world to kill fire ants and banana bugs, this DDT-like chemical was blamed for an outbreak of mysterious worker illnesses, including blurred vision, tremors, loss of memory, and unexplained nervousness.

Testifying before a Senate Agriculture subcommittee, one workman said that his hands were shaking so badly that "I couldn't drink a cup of coffee without spilling it on my clothing, and when I walked I sort of bounced." Other workers who complained to the company doctor said that they were told they were suffering from overwork or excessive drinking, and that Kepone had nothing to do with it. Meanwhile, the city government had closed the James River for fishing because it was so badly polluted from effluents emitted by Life Science Products, resulting in a loss in income to fishermen expected to run into the millions.

None of the workers was warned about the dangers of Kepone, the suit claims, even though the manufacturer was aware from research conducted in the 1960s that the pesticide caused sterility and liver cancer in laboratory animals. The report had evidently been suppressed, and it was not until recently that a study made at the Life Science plant by the American Cancer Society revealed the truth. Meanwhile, the company—which operated out of a converted service station—had gone out of business, so the workers transferred their suit to the company's contractor, Allied Chemical. Also included in the action are the state air, water, and health agencies, the E.P.A., and the Occupational Safety and Health Administration, all of whom neglected to police the chemical factory.[9]

The 1973 E.P.A. order did not affect the export of DDT to other countries, nor did it affect use of other chemicals of the chlorinated hydrocarbon group. A year after the ban went into effect, the American pesticide industry was doing business as usual to the tune of 1.2 billion dollars in domestic and foreign sales.[10] Eager to increase crop yields to feed starving masses, underdeveloped "third world" countries in Africa, the Middle East, and Asia bought America's unwanted DDT. Studies made by Dr. Chester Himel, a University of Georgia entomologist (the science of insects) "show that up to 99 percent of pesticide sprayed with current equipment is wasted, because only from one to ten percent of spray droplets are small enough to be effective."[11]

Years ago, rain water was considered a symbol of

purity, but today it is likely to be laced with a long list of pesticides that spreads an international chain of death and desolation. Most of these chemicals are wafted into the air and absorbed by clouds that carry them to every part of the world. Sprays applied in Africa have been found to fall out in rains over Europe and the Americas, all the way to the North and South Poles. Although no pesticides have been sprayed within thousands of miles of them, both Eskimos and polar wildlife have been found to contain DDT in their fatty tissues. The independent *Pesticides Monitoring Journal* which has been keeping track of pesticide residues in the American diet since 1968, reports that while DDT contamination of vegetables, grains, and fruits has diminished since the ban, there has been no significant decrease in the amount carried by cattle, other livestock, poultry, and fish. The damage has already been done, and even if DDT and its sister sprays were immediately banned in every country of the world, its residues will remain in our food for another 100 to 200 years.[12]

As Frances Moore Lappé suggests in her excellent book, *Diet for a Small Planet*, the only way we can minimize this threat to health is to "eat low on the food chain." Ms. Lappé presents a convincing case for vegetarianism when she points out that people who eat animals become "the final consumers, and thus the recipients, of the highest concentration of pesticide residues." This is because we are ingesting all the DDT that the animal has stored in its system (just as the large fish ingests the DDT stored in the body of the small fish). Those of you who feel unable to give up what the author calls "the great American steak religion," would be well-advised to substitute soybeans, legumes, and other high-protein vegetables for at least half of your weekly meat meals. Growing your own vegetables or buying them from certified organic growers further reduces the risk of body pollution and its attendant ills. Try to accustom your taste to skim milk, cottage cheese, and other dairy products low in fat and DDT.

"Nerve Gas" Pesticides—Instant Overkill

In the summer of 1969, public indignation rose to a fever pitch when it became known that the Department of Defense was planning to ship 27,000 tons of chemical war gases by rail from Colorado to New Jersey for dumping into the Atlantic Ocean. Led by Congressman Richard D. McCarthy of New York, citizens' groups denounced the Pentagon's lack of concern for the well-being of those who might be poisoned if the containers leaked or exploded in transit through densely populated areas. The Pentagon did it anyway, for they were well aware that many times that amount of organic phosphate pesticides—close relatives of nerve gases—were yearly shipped around the country by truck, and thus exposed to the even greater hazard of traffic accidents.

If the public were aware of this constant menace the outcry would undoubtedly be greater and something might finally be done to stop these shipments. But the pesticide industry maintains a low profile, which includes neglecting to mark many of their tank trucks with the possibly customary "Danger" warning. Few people are aware of the dozens of annual pesticide disasters that occur on America's highways. In her book, *Unfit for Human Consumption,* Ruth Mulvey Harmer reports that many of the accidents involve leaks and spills rather than collisions and turn-overs.

Three persons were hospitalized several years ago after an American Potash Company truck leaked nearly a ton of a nerve gas pesticide along a heavily traveled Arizona highway. Thanks to the direction of the wind and the quick thinking of the highway patrol, who closed off 60 miles of road and decontaminated it, casualties were kept to a minimum. "The U.S.D.A. reported 540 individual...poison container failures in transit during the first nine months of 1968," writes Ms. Harmer. "There is no way of knowing how many persons were inadvertently exposed to poisons as a result of those, or the other failures which went unreported."

Despite their name, organic phosphates are as synthetic as DDT. They were intended to be a "less harmful" substitute. While it is true that these artificial compounds break down in the soil in from 10 to 12 days and subsequently do not accumulate as readily in living tissue, they nevertheless present formidable dangers to human beings and the environment. They are actually *more* poisonous than DDT over the short term and have very adverse effects on most living creatures if they are inhaled or eaten in chemically treated foods.

The most potent of this group is parathion, developed from World War I nerve gases, which is used almost as indiscriminately as DDT was. Absorbed through the skin, inhaled, or ingested, parathion attacks the nervous system and inhibits the action of an enzyme (cholinesterase) necessary for sending signals to the brain. When the connection between the nerve synapses, or junctions, is severed, the nerves are kept in a state of stimulation, as if they were constantly being sent messages. Symptoms begin with a runny nose, ringing in the ears, a tendency to daydream, and may include irritability and depression; any one or all of which, of course, can be easily misdiagnosed. Extreme parathion poisoning can produce nausea and vomiting, loss of control of the bladder and bowels, drooling and sweating, and blindness. If untreated, the patient goes into convulsions, the heart, muscles, and lungs slowly become paralyzed, resulting in suffocation.

One of the most hazardous methods of applying pesticides is spraying them from a low-flying airplane, a practice that has resulted in numerous deaths of "crop-duster" pilots. One crop duster crashed outside of Belpre, Ohio, in 1968, and crawled from the wreckage of his plane through a pool of parathion that had leaked from his tanks. By evening 11 people had to be hospitalized, and three of them, including the pilot, died. A nearby dairy farm lost most of its cows and had to stop milking the survivors.

Another farmer lost all his cattle and hogs within a few minutes after he sprayed them heavily with TEPP, another organic phosphate. The farmer said that the container the chemical came in was labeled with the warning to wash with water if any of the pesticide spilled on the skin and to call a doctor if taken internally. There were no other instructions. The container also had pictures of farm animals, from which the farmer inferred that animals could be sprayed directly.

More than 100,000 pesticide poisonings are reported each year, and countless others are mistakenly treated as "flu." While people working in parathion manufacturing plants are constantly tested for indications of toxic symptoms, farm workers are not. Testifying before Senator Walter F. Mondale's Subcommittee on Migratory Labor in 1969, pesticide critic Jerome B. Gordon tried to convince legislators to correct this situation. "Uncounted thousands of the nation's migrant farm workers, farmers, and suburban homeowners have been fatally overcome or seriously disabled," Gordon protested, his appeal falling on deaf ears.

Conducting a study of the nutrition of farm children in a rural California county, Dr. Lee Mizrahi, the head of a free clinic, discovered that nearly half of the children he examined showed signs of organic phosphate poisoning. Many were listless, apathetic, with burning eyes and skin rashes, while others showed excessive vomiting and difficulty in breathing, which are signs of advanced damage. "To me, it is tragically absurd that in 1969 such a study by an obscure rural doctor should be the first one ever done on children," Dr. Mizrahi told a House Subcommittee. "We think this problem is widespread."[13]

Unable to get the Federal government to take action, Dr. Mizrahi brought his evidence to Dr. Thomas Milby, chief of the California State Department of Public Health, who compiled an official report. Dr. Milby agreed that parathion poisoning had reached "near-epidemic proportions" since replacing DDT and that an in-depth study should be undertaken. As of this writing, the insecticide is still being used.

If the tragedy that recently occurred on the island of Jamaica had happened here, it is certain that immediate action would be taken. In January, 1976, seventeen Jamaicans died and 78 were hospitalized after eating flour contaminated with parathion. Used in the spicy meat pastries that are a Jamaican speciality, the flour had absorbed the chemical from the hold of the ship that had transported it from West Germany. Not even cooking had destroyed its potency.[14] Pesticide manufacturers attributed the incident to "improper handling," rather than regarding it and similar incidents as a reason for developing an effective means of insect control that does not kill humans.

Herbicides—Turning the Earth into a Desert

Also known as defoliants, herbicides are used by

farmers to kill troublesome weeds that keep growing back
among food crops. Developed from a biological research
project conducted by the Army during World War II,
herbicides enter the plant's system, causing its metabolism
to speed up at a fantastic rate that literally makes it grow
and die overnight. Herbicides actually rearrange and
destroy the plant's chromosomes, the carriers of the genes
that direct reproductive cells, which can cause either a
defective new growth that will not survive, or a cancerous
condition that kills the plant. Obviously, the question
arises whether these chemicals will have the same effects
on man.

The U.S.D.A., the U.S. Fish and Wildlife Service, and
the F.D.A. were unconcerned about the answer, however,
for they sanctioned unlimited application in 1948, without
first conducting proper tests. These science-fiction-like
plant killers subsequently became one of the fastest
growing and most profitable branches of the pesticide
industry, and production increased by nearly 300 percent
in the years between 1966 and 1972. In the space of a few
years, herbicides equaled the pollution caused by DDT
and other pesticides when they were applied indiscriminately to ditches, pastures, crop fields, weeds on water,
forests, and even backyards.

The most popular of these, 2,4,5-T (2,4,5-Tri-
chlorophenoxyacetic acid) was used as was originally
intended and was dropped over one-eighth of the total
area of South Vietnam. Jungle vegetation was cleared and
ambush cover removed, but so were valuable croplands
and rubber plantations. In March, 1966, the State
Department gave a shocked American public the
unsubstantiated assurance that "the herbicides used are
non-toxic and not dangerous to man or animal life. The
land is not affected for future use."[15]

The South Vietnamese Government was also masking

the harm done by defoliants, which were primarily responsible for the birth of many deformed children whose mothers had breathed herbicide-laden winds that had blown over urban centers. Fearing another outburst of anti-American demonstrations, Premier Thieu forbade local newspapers to report the incident. Rumors of what had happened leaked out, however, and in 1968 the U.N.'s World Health Organization began a year-long investigation to get at the truth. Released on November 21, 1969, WHO's report cautiously stated that defoliants used in Vietnam seemed to be related to the increase in birth defects, and that their use should be immediately discontinued.

The United States Government was already aware that 2,4,5-T was a biological timebomb, however, having commissioned a study from the Bionetics Research Laboratory in Bethesda, Maryland. Among other things, Bionetics found that mice and rats given large oral doses of 2,4,5-T in the early stages of pregnancy "showed a statistically increased proportion of abnormal features" in their offspring, "in particular, cleft palate and cystic kidneys were significantly more prevalent." The report was ready in 1966, but it was not released by the government until three years later, and then only because concerned scientists threatened to bring it before the public. In the meantime, the Dow Chemical Company was selling tons of the defoliant and increasing its stock market rating while thousands of pregnant Asians inhaled 2,4,5-T from clouds spewed by American war planes, and on a less radical though no less dangerous level pregnant women in America were exposed to this lethal chemical while working on their yards.

The Bionetics report mobilized the powerful chemistry industry lobby, which immediately produced their own studies which purported to prove that experiments on animals did not give irrefutable evidence that similar results would occur in humans. This claim was rejected as nonsense by most scientists in view of what had occurred in South Vietnam. Furthermore, the deaths of millions of young chicks in a hatchery in the late 1960s was laid to residues of 2,4,5-T in corn by-products that were used in their feed. While they might be able to cast doubt on the herbicide's role in birth defects, the industry could not deny such hard evidence of its extreme toxicity.[16]

"Agent orange," as 2,4,5-T is called by the chemical industry, was finally banned by the E.P.A. in 1973, except for military use. A close chemical relative, 2,4-D, has taken its place on food crops, lawns, and golf courses, and little is known about the possible side effects of this new herbicide—unless the government is suppressing another negative report. For the safety of their unborn offspring it is recommended that pregnant women avoid all herbicides for yard use.

South Vietnam stands as a frightening example of what can happen to the land when herbicides are indiscriminately applied in massive doses. As of March, 1976, WHO reports that thousands of acres of soil are so completely sterilized there it will be years before they can once again support life. Mosquitoes cover the land and a malaria epidemic is in progress; tuberculosis bacteria have infected much of the population; and leprosy and bubonic plague are once again spreading and taking lives, going against the trend of most other countries. Radical tampering with its ecology has left the country in ruins.[17]

Mercury Poisoning—"Mad As a Hatter"

For years, most of us thought of mercury as the silvery liquid inside a thermometer. Scientists had long been fascinated by this curious metal—the only one that is liquid at room temperature—which is so dense that it will not wet skin or clothes but will penetrate the finest crevices like water. Manufacturers considered mercury to be absolutely safe to human beings, and they used it in a number of ways that most of us were not aware of. Light switches, furnace thermostats, and air conditioner switches depend upon a tiny drop of the metal to open or close a circuit; fluorescent lamps and street lights turn night into day with mercury vapor, which provides 70 percent of America's lighting[18]; floor waxes, paints, fabric softeners, camera film, wash-and-wear clothing, and the antiseptic mercurochrome all depend upon mercury to some extent; even dental fillings contain a mercury amalgam.

In 1970, farmers in the United States sprayed 800,000 pounds of mercury pesticides over millions of acres of cropland, [19] and planted millions of seeds coated with mercury to protect them from funguses. The practice might have continued to this day, if not for a tragic incident that revealed mercury's deadly double nature.

Sent home from school because she seemed dazed and unsteady on her feet, Ernestine Huckleby, an eight-year-old girl in Alamogordo, New Mexico, was promptly put to bed by her parents, who believed she was suffering from a mild case of "flu." Instead of recovering, however, Ernestine lost control of her bodily functions during the next few days and was rushed to a hospital, where she lapsed into a coma. Doctors believed the girl was afflicted with viral encephalitis—and then the mysterious malady struck again. Overnight, Ernestine's fourteen-year-old brother became paralyzed and he could see only straight ahead as if looking through a tunnel. Two days after Christmas in 1969, Ernestine's 22-year-old sister was similarly afflicted and deaf as well.

Playing a hunch that this unexplained sickness might be a form of heavy metal poisoning, New Mexico Medical Service Director Dr. Bruce Storrs sent urine samples of the three people to the F.D.A.'s toxicology laboratory in Atlanta. The report was positive; all three contained high levels of methylmercury, the most poisonous form of the metal. State and Federal health officials traced the source of the mercury to several bags of grain that Ernest Huckleby, the children's father, had found and carted home to feed to his hogs. The seeds had been dyed pink to show that they were coated with a highly toxic fungicide called Panogen. He was not going to give them to his family to eat, and Huckleby assumed that it was all right to feed these "bad seeds" to livestock.

Several of the pigs eventually died of what a health official called "the blind staggers," in which the animal has a loss of equilibrium and runs around bumping into things. They were slaughtered and the mercury-laden pork fed to the family daily for a period of three months. As a result, three Huckleby children were permanently paralyzed, and two of them became blind. The fourth, born to Mrs. Huckleby five months later, poisoned in his mother's womb, was born blind and severely retarded. Obviously, dying seeds to alert consumers that food has been coated with a deadly toxin is a totally ineffective method of warning. The real question is, of course, why contaminate them at all?

Similar "mysterious" epidemics due to mercury coated seeds had been reported previously in Iraq in 1956, in West Pakistan in 1961, in Guatemala in 1963, and again in Iraq in 1972 when villagers used the grain seed to make bread and feed animals. That year, 450 people died in Iraq and thousands more were permanently afflicted.[20]

In 1966, health officials in Sweden had warned the United States about the dangers of mercury poisoning when they banned the metal's use in agriculture and placed severe restrictions on its use in industry. The government also was aware of the mass poisoning that occurred in the Japanese village of Minamata between 1953 and 1960, where more than 100 people died or suffered severe disabilities from eating fish contaminated by mercury waste from a plastics factory. The United States seemed to be unconcerned, however, perhaps because so little was known here about mercury's effect on the food chain.

Inorganic mercury has long been recognized as a hazard to miners and factory workers, who shake so uncontrollably that they cannot walk or feed themselves when they inhale or absorb too great a quantity of the invisible, odorless fumes. The expression "mad as a hatter" originated from the mental disability exhibited by nineteenth-century hat makers, who spent the day dipping felt strips into tubs of mercuric nitrate to soften them for shaping. This type of poisoning binds mercury to body proteins, but the damage is reversible and the metal will be excreted slowly over a period of several months. On the other hand, *organic* mercury poisoning accumulates permanently in the brain and other body tissues and results in irreversible damage. The methylmercury that inflicted such tragic damage on the Huckleby family and others in various parts of the world is a member of this group.

Formerly it was believed that all mercury compounds, whether pesticide and seed residues, or effluent from chemical factories, simply sank to the bottom of oceans and waterways when they were washed from the land. Being heavier than water and of no food value to ocean life, the metal was thought to stay there for thousands of years until it was finally broken down and reabsorbed into the earth. Three researchers, Dr. Carl Rosen of Sweden,

Dr. J. M. Wood, and F. S. Kennedy—both from the University of Illinois—knew otherwise and brought their findings to the attention of U.S. officials in 1967.

The Rosen-Wood-Kennedy research team had discovered that micro-organisms in the mud of seabeds were able to absorb organic mercury and turn inorganic mercury into the more poisonous methylmercury. Algae subsequently ate the metal-contaminated bacteria, small fish ate the algae, and large fish ate the small fish. The poison was subsequently concentrated many hundreds of times over in the largest fish—along with massive quantities of DDT and other pesticides—which meant that the greatest danger lay in eating tuna and swordfish, as they swim closest to shore and their fat tissue absorbs the poison most readily.

Spurred to action by the Huckleby tragedy, the Rosen-Wood-Kennedy findings, and the coverage given in the media, the F.D.A. imposed a tolerance level of 0.5 parts per million of mercury in fish in 1970, three years after the evidence was reported. One million cans of contaminated tuna were quickly removed from grocers' shelves, prompting many to wonder how many poisoned cans had been consumed by Americans in the 30 years or so that mercury has been dumped into the oceans. Swordfish was found to contain as much as 50 times the acceptable level of mercury, and it has virtually disappeared from the market.

On February 18, 1976, the Environmental Protection Agency finally banned the production of virtually all mercury pesticides. "Economic, social and environmental costs and benefits of the continued use of mercurial pesticides are not sufficient to outweigh the risk to man or the environment," proclaimed Russell E. Train, new chief of the agency. Fungicides containing mercury compounds to treat awnings intended for outdoor use, and those used for the control of Dutch Elm disease and brown mold were not affected.

While it is unlikely that the ban will result in decontamination of fish in the near future (possibly not within our lifetimes), you can minimize the risk by eating only small fish, which do not contain as many pollutants. If tuna is your dish, buy only the albacore, which weighs about 45 pounds and has white-colored flesh. Yellow fin—the tuna that was removed from the market—weighs about 150 pounds, and big eye and blue fin average about 230 pounds, making them poor risks. Remember that even though you can't see, taste, or smell mercury and DDT in big fish, these contaminants are there just the same.

Natural Alternatives to Pesticides

Mounting evidence indicates that the indiscriminate use of pesticides has actually increased the number of pests while lowering the quality of human life. Malaria-carrying mosquitoes are once again on the rise, and many other species of insects have mutated and developed a genetic resistence to man's chemicals. In California, cotton growers who sprayed their fields to get rid of the lygus bug discovered that the pesticide also killed predators of the boll worm, when then began to increase. The spraying had actually brought back an old pest.

Farmers in the Cañete valley of Peru had been so impressed by the efficacy of DDT that they sprayed it over their fields "like a blanket." Four years later, the pests returned with a vengeance, along with five new insects. Their natural enemies had been killed and these new predators quickly took over the land and decreased the yield to pre-pesticide levels. The list of similar incidents is endless.

While these "super-flies" may have outwitted chemists, they cannot outwit nature. The best way to combat crop pests is to develop and grow species of grain, fruits, and vegetables that have a natural resistance to insects, mites, and funguses, and to institute a program of biological controls. A new strain of alfalfa was developed in the 1960s which effectively resists the dread spotted alfalfa aphid that once nearly destroyed the entire alfalfa industry. Illinois biologists stopped the spread of the western corn rootworm, which had developed an almost total resistance to pesticides, by yearly alternating corn and soybean crops in the same fields. Rotating the crops solved the problem because the worm does not eat the soybean plant and dies when soybeans are planted The screwworm, which used to kill millions of dollars worth of southern cattle each year, was eliminated through the introduction of sterile strains of the insect that produced no offspring at breeding time. Other pest damage can be kept to a minimum by rigorous tilling of the soil to destroy insect larvae, and staggering crops so that no single insect problem gets out of hand.

If these methods are not enough, *botanical* pesticides can be used. These come from plants, and most are safe to warm-blooded animals with a few exceptions, such as the highly poisonous nicotine extract. However, pyrethrum, obtained from two species of chrysanthemums, is not environmentally disruptive, nor is rotenone, which comes from the root of the derris plant. In buying a home insecticide always look for products that contain only these natural extracts, and never buy a mist spray "bug bomb." While they are organic, pyrethrum and rotenone can nevertheless cause skin and lung irritations.

Most of these biological, non-chemical, and environmental methods of insect control were known as far back as the turn of the century, yet they account for only 10 to 20 percent of pest eradication activities.[22] Chemical pesticides may be cheaper and an easy shortcut but their long-term residual effects are proven to be dangerous and ultimately very expensive.

The Future of Pesticides

Meeting under the aegis of the National Academy of Sciences in Washington, D.C., early in 1976, sixty specialists who had worked independently in pesticide use and hazards to health issued the following statement:

"We believe the state of Federal knowledge in this critical area is highly inadequate. The pest control enterprise places a billion pounds of toxic materials into the environment each year, but it is 'normal' for us to have only the vaguest idea of how much of each compound was used and where, and even then only after half a decade's lag."

Preparation of their report, which represents the viewpoint of the American science establishment, took a total of three years. During this time, investigators were surprised to learn what many conservationists knew all along: Injuries from pesticides are seriously underestimated because there is no effective reporting or diagnostic system. The panel called on the government to institute programs to identify and study this menace to public health, and to develop alternatives to chemical pesticides.[23]

Although this is a step in the right direction, new legislation is needed if the chemical industry is to be effectively censored and prevented from developing and marketing chemical substances which are harmful if not fatal. At present there are simply not enough F.D.A. and E.P.A. inspectors to assure that set tolerances and bans are adhered to. However, no legislation can eliminate or even reverse the damage already done. But in the absence of new restrictive laws, the pollution of our bodies by past, present, and future pesticide sprays will continue unchecked.

Notes

1. *Chemical Hazards in the Human Environment,* Paper from 11th Science Writers Seminar, American Cancer Society, New Orleans, La. March 28, 1969.
2. *The New York Times,* February 6, 1976.
3. Trager, James, *The Food Book,* Avon Books, New York, 1972.
4. Marine, Gene and Van Allen, Judith, *Food Pollution: The Violation of Our Inner Ecology,* Holt, Rinehart and Winston, N.Y., 1972.
5. Lappé, Frances Moore, *Diet for a Small Planet,* Ballantine Books. New York, 1976.
6. Dr. Hargraves testifying on *Dangers of Lindane,* Ribicoff Hearings, Part 2, pp. 484-97.
7. Turner, James S., *The Chemical Feast,* Grossman Publishers, New York, 1970.
8. "The Effect of Known Repeated Oral Doses of Chlorophenothane, (DDT) in Man," *Journal of Am. Med. Assoc.,* October 27, 1958.
9. *The New York Times,* January 28, 1976.
10. U.S. Bureau of the Census, Annual Survey of Manufacturers, 1973.
11. *The Christian Science Monitor,* December 11, 1969.
12. "Systems Studies of DDT Transport," *Science,* 170:503-8, 1970.
13. Dr. Mizrahi, Lee, Testimony, House Subcommittee on Labor in San Francisco, November 24, 1969.
14. *The New York Times,* January 30, 1976.
15. Whiteside, Thomas, "Defoliation" *The New Yorker,* February 7, 1960. p. 33.
16. *Ibid.*
17. *The New York Times,* March 21, 1976.

18. *The Statistical Abstract of the United States*, 1976.
19. Montague, Peter and Katharine, "Mercury: How Much Are We Eating?," *The Saturday Review*, February 6, 1971.
20. Putman, John J., "Quicksilver and Slow Death," *The National Geographic*, October, 1972.
21. *The New York Times*, February 18, 1976.
22. "Alternate Methods of Controlling Plant Pests," *F.D.A. Papers*, February, 1969, p. 16.
23. *The New York Times*, February 6, 1976.

Food Additives

By now the reader is well aware that just about all the food we eat has been chemically treated at each stage of its journey from the feedlot or field to the supermarket shelf. Whether of animal or plant origin, its growth pattern, size, appearance, texture, and nutritional value have all been manipulated and transformed by a plethora of chemicals, many of which may be hazardous to our health. In this chapter we will take a look at the various chemicals that make their way into our food during the final stage of commercial production—during the processing and packaging. These chemicals, which we shall classify under the general heading of "food additives," range from reasonably benign food colorings, to such highly dangerous substances as sodium nitrate and sodium nitrite.

BHT and BHA

BHT (butylated hydroxytoluene) and BHA (butylated hydroxyanisole) are petroleum products used widely as antioxidants in foods and other items. These fat-stabilizing substances were introduced to replace the natural oxidants which are lost during the commercial processing of refined fats and oils.

First used mainly in fats and fat products, BHT and BHA are now found in almost every factory processed food, and even in many packaging materials. Some of the many items which contain BHT and/or BHA are lard, chicken fat, butter, cream, shortening, bacon, potato chips, processed meat and fish, pastries, cakes, candies, peanut butter, nutmeats, raisins, milk, imitation fruit drinks, breakfast cereals, spices, and pet foods. BHT and BHA are found in containers used for milk, ice cream, cottage cheese, cereals, potato chips, and cookies. They are also added to chewing gum, drugs, and cosmetics.

BHT and BHA first enter the meat we eat in the animals' feed. Many processors of livestock feed use these antioxidants to stabilize the vitamins in the animal fats used to fortify the feed, to improve the taste and appearance of the feed, to control dust and reduce the risk of fire, and to make handling and shipping of the feed easier.

One would assume that any substance or substances in such wide use have been conclusively tested for safety. This, however, is not the case. The story of one group's attempts to find out exactly what tests had been conducted on these antioxidants shed a harsh light on the reliability of the Food and Drug Administration, the government agency responsible for the safety of the products we consume.

In the early 1960s, persons cooperating with Consumers' Research wrote to the F.D.A., requesting information on the tests used to establish the safety of BHT and BHA. The F.D.A. supplied these individuals with four references, only three of which could be located in libraries. This fourth reference was a report issued by the American Meat Institute, who claimed that no copies of the report were available. The F.D.A. said its own copy could not be made available, either.

The Meat Inspection Division of the United States Department of Agriculture, to whom the report had been submitted,revealed that the toxicity studies in question had been conducted by private companies rather than by a government agency. They said they were unable to make public the information, as it had been submitted to them "in confidence." One wonders what kind of perhaps questionable information might be contained in a report that must be kept from the public's eyes.

The F.D.A. sent copies of only one study to the individuals who had requisitioned information. This study, which was published in 1955, was financed by three manufacturers of BHT. In the experiment, four species of animals were fed various dosages of BHT. The experimenters determined what level would kill the animals in a single dosage, and what lower level could be tolerated by the animals over a two-year period. Although some animals on the lower dosage died, their deaths appeared to be unrelated to the BHT. Some of the control animals, who had received no BHT, also died. Based on this conflicting, limited evidence, the researchers concluded that BHT had not caused any of the deaths. This experiment appears to have been the basis of the Federal agencies' approval of the use of BHT in foods.

The F.D.A. failed to mention other, earlier studies that had raised serious questions about the safety of BHT. Experiments conducted in Romania in the 1950s revealed metabolic stress in rats fed BHT. Eventually the Romanian Hygiene and Public Health Institute recommended that BHT be banned from use in foods.

The Australian researchers W. D. Brown, A. R. Johnson, and M. W. O'Halloran found in 1959 that BHT in combination with lard reduced the growth rate and weight of male rats. The reduced growth rate was directly attributable to the BHT. They also found that BHT caused an increase in the weight of the liver; this was suggested to be the result of the extra stress placed on the liver, which is the organ responsible for the detoxification of toxic substances.

Since 1959, sporadic experiments on BHT and BHA

have been conducted. Most of them have not originated in the United States, whose population as a whole is directly affected by the use of these chemicals, but in Europe, and in Australia and New Zealand. No conclusive evidence on the safety of BHT and BHA has yet been reached. However, this fact should be cause for further study, *not* for complacent acceptance by Federal agencies of the rampant use of these chemicals.

There have been some reports of possible beneficial effects of BHT and BHA. Drs. R. J. Shamberger, S. Tytko, and C. E. Wills, of the Cleveland Clinic, reported in 1972 on the possible connection between the use of antioxidants and the decline in the number of deaths from stomach cancer. They point out that stomach cancer used to be the leading cancer cause of death among men in the United States. Since 1930, the incidence and mortality rates of stomach cancer have declined steadily; stomach cancer is now the fifth leading cancer cause of death among men, and the eighth among women. It was in the early 1930s that breakfast cereals (which contain some natural antioxidants) began to gain wide acceptance in the United States, and the late 1940s that chemical antioxidants first came into use. The decline in stomach cancer which began in 1930 accelerated significantly in the late 1940s. The researchers also point out that there has not been a similar decline in the rate of stomach cancer in Europe, where breakfast cereals are much less used.

Some individuals have been found to be sensitive or allergic to BHT and BHA. Reactions to these chemicals have included skin blisters, hemorrhaging of the eye, weakness, edema, and discomfort in breathing.

Unfortunately, it is not only difficult to detect the cause of an allergic reaction in this age of chemical additives, but also difficult to ascertain what chemicals any one product contains. A label may merely list the inclusion of a "freshness preserver," or say that there is an "antioxidant added." Such "information" is of little use to the person seeking to avoid BHT and/or BHA.

Apparently, officials in Great Britain take the evidence against chemical antioxidants more seriously than do officials in the United States. In Great Britain limited use of BHT is permitted in lard, butter, margarine, and essential oils. In 1958, the British Food Standards Committee recommended that BHT be taken off the list of permitted antioxidants. Due to protests from industry, their recommendation was not acted upon. BHT was banned, however, from baby foods—a step which had been requested by the British Industrial Biological Research Association.

The reader is urged to protest the widespread use of these chemical antioxidants. Letters sent to the processor of a breakfast cereal and to a processor of wheat germ led to their decision to discontinue use of BHT and BHA in the two products.

Sodium Nitrite and Sodium Nitrate

Sodium nitrite and sodium nitrate are used extensively in the curing of meats and smoking of fish. Nitrites are most often used (although nitrates were used originally); nitrate itself is not actually a curing agent, but serves as a source of nitrite. The reader who makes a quick survey of the packaged meat shelf of any local supermarket will be astounded (and after reading this section, horrified) to see how many products contain sodium nitrite.

The toxic factor in nitrates from vegetables, especially spinach, can cause a relatively rare disease called methemoglobinemia, to which infants are particularly susceptible. When this condition occurs, the blood's hemoglobin, which carries oxygen to the tissues of the body, is changed into a form called methemoglobin. As methemoglobin is unable to transport oxygen, the person turns blue and becomes dizzy. In extreme cases, death may result.

The primary concern among scientists and consumer advocates is the relation between sodium nitrite and cancer. Although nitrite by itself has not been shown to cause cancer in test animals, it participates in a chemical reaction that results in a particularly potent cancer-causing substance. When the nitrite combines with certain amines (amines are a product of protein breakdown, and are also found in certain drugs and other non-food substances) in a mildly acidic environment such as the human stomach, it forms nitrous acid. The nitrous acid then reacts with the amines to form nitrosamines; nitrosamines are among the most potent carcinogens (cancer-producing substances) yet discovered by scientists.

One of the most alarming characteristics of nitrosamines is that they do not attack only one organ of the system, such as the lung, breast, or stomach; they appear to be able to produce cancer in any or all parts of the body simultaneously. In addition, no species of animals has been found to be resistant to them. They have produced cancer in hamsters, mice, rats, dogs, guinea pigs, and monkeys. So there is no reason to think that the human species is exempt.

Nitrosamines were first discovered to be cancer-producing in 1956 by British scientists. In 1963, a German chemist put forth the suggestion that nitrosamines might be produced in the stomach by a chemical reaction involving nitrite. Although some members of the F.D.A. question whether harmful nitrosamines can actually form in the stomach when "normal" levels of nitrite and amines are consumed, the evidence against the use of any level of nitrite has become overwhelming.

About 80 percent of the nitrosamines tested in laboratories have been proven to be carcinogenic in animals. Some have produced cancer at extremely low dosages, while others have produced cancer after only a single dose. In some cases, nitrosamines given to female animals toward the end of pregnancy have left the mother unaffected, but caused the young to later develop cancer of the liver, kidney, brain, and spinal cord.

Unfortunately, the human stomach appears to be the ideal environment for the production of nitrosamines. When cats and rabbits have been fed as little as two hundred parts per million of nitrite (the amount allowed in meat and fish) along with the amine diethylamine, the nitrosamine diethylnitrosamine was produced in their stomachs. This is especially significant, as the stomachs of cats and rabbits are of approximately the same acidity as the human stomach. The acidity level affects some control over the rate of nitrosamine formation.

In other experiments, in Germany, sodium nitrate and an amine were given to 31 human volunteers. A nitrosamine (not believed to be carcinogenic) was found to have formed in their stomachs. This is further proof that nitrosamines can indeed form in the human stomach.

One of the foremost scientists studying the effects of nitrites is Dr. William Lijinsky, of the Oak Ridge National Laboratory in Tennessee. In one series of experiments, Dr. Lijinsky fed nitrites and amines regularly to test animals over a period of time. The animals developed the same kind of cancer as did those animals fed the corresponding nitrosamine. This is virtually indisputable proof that the dangerous nitrosamines are able to be produced in the stomach when nitrite and amines are consumed.

In one of Dr. Lijinsky's experiments, a group of rats were fed nitrite and an amine found in a particular drug. Within six months, 100 percent of the rats had developed malignant tumors. Normally, if a substance is found to produce cancer in only 50 percent of a group of test animals it is considered to be a potent carcinogen.

Thus there seems to be little question that when nitrite and amines are present in the stomach at the same time, nitrosamines can form—and that most of these nitrosamines could be highly carcinogenic. Obviously, we can do nothing about the fact that our stomach provides the ideal acidic environment for nitrosamine formation. Amines, too, are relatively difficult to avoid (only the amines called secondary and tertiary amines react with nitrite to form nitrosamines). Nitrosamine-producing amines are found in wine, beer, tea, cereals, fish, cigarette smoke, and numerous drugs. Among these drugs are anesthetics, tranquilizers, diuretics, antihistamines, antidepressants, oral contraceptives, and high-blood pressure medications.

Nitrite is, or should be, the easiest to avoid of the three factors needed to produce nitrosamines. Small amounts in some water supplies and certain vegetables are difficult to avoid—but the vast majority of the nitrites we ingest are those added needlessly to meats and fish.

In addition to the likelihood of the nitrite in meat reacting with amines in the stomach to form dangerous nitrosamines, there is the ever-present risk of nitrosamines forming in the meat itself. In 1972, the F.D.A. and the Agriculture Department found nitrosamines in eight samples of processed meat ready for marketing. Dried beef and cured pork contained 11-48 parts per billion, ham contained 5 parts per billion, and hot dogs contained 80 parts per billion. Four brands of bacon which were free of nitrosamines when raw were found to contain 106 parts per billion when cooked, and twice that amount were found in the bacon drippings. The F.D.A. has also found up to 26 parts per billion of nitrosamines in smoked chub and salmon.

With such mounting evidence against nitrites, one might wonder why their use continues to be permitted. The story of the debate over whether to ban nitrites is one of economic interests and politics. Apparently, the health of our country's citizens is the last thing on the list to be considered.

Many members of the meat industry and of the government claim that nitrites serve as a necessary protection against botulism, the often fatal form of food poisoning caused by the germ *clostridium botulinum*. According to the United States Department of Agriculture, nitrites damage the spores (reproductive cells) of the bacteria during cooking. Thus, if the bacteria later break out of their spores, they are unable to grow or reproduce.

At first glance, this argument may appear to make good sense. However, botulism from meat is exceedingly rare—in fact, there is no recorded case of human botulism caused by cured meat. And if nitrites are necessary to protect against botulism, why hasn't the U.S.D.A. set a minimum required level of nitrites for all cured meat?

The F.D.A. is a little more honest about nitrites and botulism. In August of 1970, they admitted in a Status Report on the chemistry and toxicology of nitrites that sodium chloride is actually the main preservative in cured meat, and that sodium nitrite is used primarily as a color fixative in meat and fish. When used without salt, nitrite is not even effective against botulism.

The level of nitrite permitted in meat far exceeds the level that would be needed to control botulism were botulism really the main concern. Astounding as it may seem, the U.S.D.A. has never actually used any scientific means to determine what level of nitrites (if any) is safe for human consumption. In 1926, they decided upon the level of nitrite to be allowed in meat by simply taking the highest level of nitrite commonly found in ham after it had been cured; this level was then made the standard.

A convenient loophole in labeling requirements allows meat processors to use nitrite in some products without listing it on the label. The U.S.D.A. does not require labels to include the color fixatives used in standardized meat products; a standardized list of these products is found in the *Federal Register*. At the same time, the U.S.D.A. defends the use of nitrites as a necessary preservative against botulism! Yet under the law, no preservatives are exempt from labeling requirements. Thus the U.S.D.A. contradicts itself. If nitrites are a preservative, they must be labeled as such; and if they are a color fixative, then their use cannot be justified as a preservative against botulism.

Although the U.S.D.A. now claims that nitrites have been necessary all along as a preventative against botulism, nitrites have received U.S.D.A. approval only as a color fixative. Thus, the use of nitrite to control botulism requires approval, which has never been obtained. The U.S.D.A. is actually committing a legal violation in allowing the chemical to be used for botulism control. Legally, the U.S.D.A. must formally propose such use of nitrite to the F.D.A., and then submit the proposal to a hearing. Naturally, the U.S.D.A. wishes to avoid such a public (and publicized) confrontation on this issue, which has already drawn the attention of consumers and consumer advocates across the country.

In 1972, Dale Hattis, a graduate student at the Stanford University School of Medicine, filed a freedom-of-information suit in order to gain access to all of the F.D.A.'s documents concerning its approval of the use of nitrite in smoked fish. In his report, entitled "The F.D.A. and Nitrite," Hattis showed the various economic pressures which had influenced the F.D.A., causing it to approve the use of a chemical it had previously classified as a harmful substance. In 1948, the F.D.A. had stated in an information letter:

"We regard [nitrite and nitrate] as poisonous and deleterious substances not required in the manufacture of any food subject to the jurisdiction of the Food, Drug and Cosmetic Act, and, as such, any food subject to the act and containing any quantity of these chemicals would be deemed to be adulterated, under the law, regardless of labeling."

Although the use of a "poisonous or deleterious" substance is legally permitted only if it is required in the production of a food, or cannot be avoided in good manufacturing processing, in 1960 the F.D.A. began to grant permission for manufacturers to use nitrite and nitrate for solely cosmetic purposes—as a color fixative in smoked tuna, smoked and cured salmon, shad, and sable fish.

In 1963, several persons died of botulism after eating smoked whitefish that had undergone either improper refrigeration or improper heating while being processed. The ensuing botulism scare provided the fish industry with an excuse to request permission for further use of nitrites. The industry justified its request by stating that small processing plants frequently lacked the appropriate facilities to heat the fish at a sufficiently high temperature, or long enough to kill the botulism bacteria. This, of course, was a request to use nitrite as a substitute for complying with the safety guidelines recommended by the F.D.A. for preventing botulism! Canada, and the states of Minnesota and Michigan, passed legislation forbidding the use of nitrite as a substitute for proper heating during the processing of smoked fish.

As was pointed out in a report prepared by Congress-person Fountain's Intergovernmental Relations Subcommittee, the American food industry certainly has the technological sophistication to protect consumers from botulism without having to resort to the use of dangerous chemicals. In fact, nitrite-free cured meats have already been produced.

Food Coloring

Over 90 percent of the colorings used by food manufacturers are synthetic; most of them are coal-tar derivatives. Food colorings are added to nearly every kind of food and beverage found in the supermarket, including meats, soft drinks, wines, bread, cakes, cereal and fruit. Manufacturers wish to create the illusion of quality by giving a food the color they think the consumer associates with quality. Florida oranges, for example, are dyed orange although in certain seasons their outside remains a spotty green and brown even when they are fully ripened.

Many consumers are quite understandably misled by the term "U.S. certified artificial color"; they assume the coloring has been certified for safety. This, however, is unfortunately not the case. A color which is "certified" has merely met certain government standards which allow the presence of an established percentage of impurities, such as arsenic, etc.

Even when a color is "delisted" (i.e., decertified), it is not completely banned—it has simply lost its certification. It may continue to be passed off onto the public in products still in storage in warehouses and supermarkets, and in products which are shipped only within states, and thus do not fall under the control of the Federal agencies. Some of the food colorings which have been delisted are Sudan I; Butter Yellow; Yellow No. 1, 3 and 4; Red No. 32; Orange No. 1 and 2; and Red No. 1 and 4.

The story of Red dye No. 2 will be examined in detail in the next section. However, the attitude of various apologists for the food industry toward this coloring is relevant to a discussion of all the colorings (and all food additives in general). Fred Stare, the author of a syndicated health column, recently stated on Barry Farber's show in New York City that as all food additives, including Red dye No. 2, are harmless, they should be used. When Farber asked if he wouldn't care to be more cautious and qualify his statement, Stare irresponsibly replied that since the body is composed of chemicals anyway, it made no difference if more chemicals were added.

The person who had co-authored a book with Stare said that Red dye No. 2 had been banned by the F.D.A. for political reasons. This reaction is typical of the way representatives of the food industry attempt to manipulate and distort the public's information on food matters. When any additive is approved by the F.D.A., even if serious evidence against it has been found, they point to F.D.A. approval as proof of its safety. Yet when the F.D.A. finally gets around to banning a substance, they immediately say that the F.D.A. only acted out of political reasons.

An ironic note is that even dog foods are colored—although dogs are color blind! That artificially produced red meat color is meant to attract the pet owners, and their dollars, not the animals that actually eat the food.

The Coal Tar Dyes

When heated in the absence of air, coal is converted into coke, coal gas, and coal tar, which is a viscous black liquid.

Approximately 95 percent of the synthetic colorings used in the United States are coal tar derivatives. Red dye No. 2 is the most infamous of these.

Violet No. 1—The safety of coal tar dye Violet No. 1 has been questioned. Violet No. 1 is used to stamp the symbol of the Department of Agriculture's inspection on meat, and as a coloring in beverages, candies, and pet foods.

In two separate studies using male rats, Dr. Lloyd Hazleton and Dr. O. Garth Fitzhugh found Violet No. 1 to be safe. A third study, by Dr. W. A. Mannell and his associates at Canada's Department of National Health and Welfare, used male and female rats, and higher dosages of the dye. Of 30 rats, five developed malignant tumors. Only one rat of the control group developed a tumor. The F.D.A. dismissed the Canadian study on the grounds that the dye used may not have fit U.S. specifications for Violet No. 1. As the dye had been used up, and the manufacturer of the dye gone out of business, there was no way to determine whether the dye was at all dissimilar to U.S. Violet No. 1.

In several F.D.A. tests, the dye was found to cause skin lesions. In 1971, the F.D.A. asked the National Academy of Sciences to evaluate all the available material on Violet No. 1. The evaluating committee dismissed the Canadian studies, and declared the dye safe. However, it also recommended that a lifetime feeding study on dogs be carried out. This in effect postponed a definitive decision on the dye until 1979 at the earliest.

Citrus Red No. 2—The safety of this dye has also been questioned. This is the coloring which is used by Florida growers to cover up the green skins of their oranges, temple oranges, and tangelos, primarily from October through December. Tests have indicated that the dye may be carcinogenic. In 1969, the FAO/WHO Expert Committee recommended that it not be used as a food coloring. At least eleven states, and Canada, have since banned the sale of artificially colored oranges.

Unfortunately, artificially colored oranges are not marked as such. Although retailers are legally required to post a notice stating "artificially colored oranges" when they sell these oranges, very few comply. Consumers should thus be cautious about eating the peels of these fruits, or using them to make marmalade.

Other Coal Tar Dyes—Tests conducted so far on Blue No. 1 and 2, and on Green No. 3, have found them to be safe. There is question about the safety of Orange B, as it is similar in chemical structure to Red No. 2. This dye is used mainly to coat the outer surface of frankfurters (which, of course, also contain sodium nitrite).

Red No. 3 has also been found to be safe. There is question whether large amounts of Red No. 4, which is used in maraschino cherries, causes atrophying of part of the adrenal cortex and changes in the bladder.

Red No. 40 was developed by the Allied Chemical Corporation as a substitute for Red No. 4. It has been approved only since 1971, and it is not known how well it has been tested (certainly no long-term studies could have been conducted).

Yellow No. 5 has so far been found to be safe. There is some indication that Yellow No. 6 may affect the eye when ingested in large quantities. Further studies must be conducted.

It must be emphasized that there is really no justifiable reason for the use of any artificial colorings in foods. When we say that a coloring has so far been found to be safe, this is merely an indication that it has not been found to be harmful. Yet it is too early to tell with any of the food additives. We do not know what long-range effects may later show up.

Red Dye No. 2

Although Red dye No. 2 was recently finally banned for use in bacon, its history is well worth examining for two reasons. First of all, it presents a clear and dismaying example of how far the F.D.A. and the food industry will go before they are forced to ban an obviously hazardous substance. Secondly, the facts on Red dye No. 2 should serve as a warning in terms of other, similar dyes—especially Red dye No. 4, which it now appears may be no safer than Red dye No. 2.

In its pure state, Red dye No. 2—or amaranth, as it is also called—is a reddish-brown powder which turns deep red when in a solution. Until banned, it was used alone and in combination with other dyes to color a vast range of the foods we eat. Products which have contained Red dye No. 2 include ice cream, ice milk, luncheon meats, frankfurters, processed cheese, fish fillet, cornflakes, shredded wheat, rolls, pretzels, cookies, cake mixes, pickles, fruit juice, canned fruit, salad dressings, jam, candy bars, and soft drinks. It was also used in a number of cosmetics, primarily lipstick; and in various drugs, including vitamins prescribed for pregnant women.

Red dye No. 2 was suspected of causing cancer as far back as 1956, when it was discussed by scientists at a symposium in Rome. In 1964, a joint FAO/WHO Expert Committee on Food Additives passed Red dye No. 2 for use in food, but added that it must undergo further research without delay.

In 1970, the F.D.A. was presented with translations of two studies on Red dye No. 2 that had been conducted by Russian scientists. One study showed that extremely small amounts of the dye caused birth defects, stillbirths, fetal deaths, and sterility in rats. Now there was strong evidence that the dye was linked with both cancer and reproductive damage.

The Russians had found that rat fetuses were endangered by only 15 milligrams of dye per kilogram of body weight—the precise amount set by the World Health Organization as safe for human consumption! The fetuses of the rats fed the dye were resorbed, which is similar to miscarriage, or spontaneous abortion in humans.

The F.D.A. dismissed these Russian experiments as invalid on the grounds that none of the control animals developed tumors; normally some cancer is found in control animals as they age. However, the F.D.A. has files on other experiments in which there were no tumors in the control animals. And regardless of any weaknesses in the Russian experiments, the F.D.A. certainly should have taken such significant findings seriously.

To the press, the F.D.A. claimed that there was no cause for worry, that Red dye No. 2 was above suspicion. They quoted experiments they had conducted in their own laboratories in the 1950s to support this claim. Yet one of their most widely quoted experiments reveals several serious inconsistencies. For one thing, Dr. A. A. Nelson, a pathologist who had worked on the experiment, had actually found tumors in rats fed Red No. 2, as well as Yellow No. 5 and 6.

Equally odd was the fact that Dr. Nelson found that there were seven rats that had never reached the pathology laboratory to be dissected. All seven of these rats had tumors, and all but one of them had been fed a dye. At this point in the mix-up, it was impossible to determine which animal had been fed which dye. Yet despite this carelessness, the F.D.A. used data such as this to tell the public that there was no cause to worry about Red dye No. 2.

Because there are no strong legal requirements for the banning of substances which cause reproductive damage (as there are for cancer-causing agents), the F.D.A. felt

more inclined to study Red dye No. 2's effects on
reproduction than on cancer. In 1971, studies using
chickens were carried out. The dye was found to cause
stunted growth, skeletal deformities, and, in some cases,
malformed eyes. Under certain conditions, most of the
chicks failed to live long enough to hatch.

In September, 1971, the F.D.A. went so far as to
announce that as of January, 1972, it intended to severely
lower the permitted level of Red dye No. 2. This would in
effect constitute a ban, as the amount allowed would be
too low to be of any use to the food industry.

Two weeks after this announcement had appeared, the
industry's Institute of Food Technologists Committee on
Nutrition and Food Safety requested that the F.D.A.
secure an objective opinion from some group such as the
National Academy of Sciences. Not only did the food
industry want to "save" Red dye no. 2, but it was also
afraid that once one chemical was banned for causing
birth defects and spontaneous abortions, other chemicals
might suffer a similar fate.

The F.D.A. eventually succumbed to the pressure of
industry and handed the matter over to the National
Academy of Sciences-National Research Council. The
scientists at the F.D.A. felt betrayed—for what was the use
of conducting experiments if the results were simply
dismissed?

Dr. Philip Handler, the president of the National
Academy, at first refused the job of evaluating the dye. He
claimed that the F.D.A. itself was fully competent to carry
out the task. For some unknown reason, however, he
changed his mind and established a committee to come to
a decision on Red dye No. 2.

Testimony from researchers was heard by the committee in February, 1972. The committee members did not

appear to take the evidence on chicken embryos too seriously. One committee member even had the outrageous bad taste to refer to Red dye No. 2's apparent abortive effect as a possible birth control pill.

After hearing reports on various industry sponsored studies, the committee decided that it was unnecessary and premature to restrict the use of Red dye No. 2. At first, the Academy's review board refused to put its name to the study. However, as the review board was only advisory, the report was eventually released three months later. The report was widely criticized. *Consumer Reports*, for example, noted that in mentioning one study, the committee referred to the part that showed no resorptions when the dye was part of the diet, while ignoring the part that showed much more negative results when the dye was fed directly into the stomach via a tube.

The F.D.A. was in an awkward situation, as the report of the Academy committee was so obviously biased. In order to save face, the F.D.A. handled the situation in the following manner: They accepted their own rat studies, which set a safe level of the dye (the level at which no embryos were damaged) at 15 milligrams per kilogram of body weight per day. According to their usual procedure, however, they should then have divided this figure by 100, to arrive at the "safe" level for human consumption. For most persons, however, this level would allow less than one bottle of cherry soda a day—virtually a ban on the dye. This radical departure from the F.D.A.'s usual hundredfold safety margin rendered all the debate and research meaningless.

Another experiment—which lasted 2 1/2 years and cost $100,000—was embarked on by the F.D.A. in 1971. This experiment was harshly criticized by various scientists. Because of careless supervision, some of the rats in the study (approximately 50 out of the 500) were mixed up and fed wrong doses of the dye.

Due to the undeniable evidence that Red dye No. 2 is hazardous to the health, and to mounting public pressure, the dye was finally banned in January, 1976. This was a hard-earned victory for the consumer, and hopefully a sign of similar victories to come.

Other Food Additives

In writing this chapter, we felt that the reader would benefit most from a detailed examination of some of the worst offenders among the many additives that go into our food, such as sodium nitrite and sodium nitrate, and BHT and BHA. Yet many more additives are used during the manufacturing process. The National Academy of Sciences publishes a book called *Chemicals Used in Food Processing*, which contains almost 300 pages of lists of chemical additives. There are approximately 3,000 food additives covered by formal regulations. Naturally, we could not cover them all in the space of this book. In this section, we will describe some of the more commonly used ones that may pose a threat to our health.

Artificial flavorings—Artificial flavorings are actually a very large group of additives. In fact, two thirds of the food additives in use in the United States are either natural or synthetic flavorings. These flavorings enable the manufacturer to skimp on the more expensive, genuine ingredients.

Regulations governing flavorings tend to be lenient. One reason for this is that flavorings are generally used in very small amounts. Nonetheless, this does not rule out a flavoring's being hazardous to the health. Safrole, which was used as a flavoring in root beer until 1960, was found to cause cancer of the liver.

Food companies are allowed to declare that a chemical is "generally recognized as safe" (GRAS). Although flavorings are not actually classified as GRAS, they are handled in a similar manner: a manufacturer can declare a flavoring safe on the basis of very little testing. The manufacturer does not even have to list flavorings specifically on the label. Practically anything could be listed as an "artificial flavoring."

Benzoyl peroxide—This is a chemical which bleaches, but does not age, flour; it is used in combination with various aging agents, such as azodicarbonamide, iodate, and bromate. Most of it decomposes to benzoic acid, which is not believed to be hazardous to the health. However, benzoyl peroxide does destroy vitamin E. This is important to know if one insists on eating products made from bleached flour.

Brominated vegetable oil—Brominated vegetable oil, or BVO, is vegetable oil which has been combined with bromine. Flavoring oils are added to BVO, and the solution is then added to fruit-flavored drinks, both carbonated and noncarbonated. The bromine, which increases the density of the oils, keeps them dispersed throughout the drink—otherwise they would float to the top of the drink.

Manufacturers are not required to list BVO on labels. Some of the drinks which contain BVO are Orange Fanta, Fresca, Patio, Orange Crush, and Mountain Dew.

Tests have shown that rats fed BVO suffered damage of the heart, liver, thyroid, testicle, and kidney. Studies conducted in England have found that fragments of BVO accumulate in animal tissue. Research must be done to determine whether BVO is able to cause cancer and/or reproductive damage.

Diethyl pyrocarbonate—Diethyl pyrocarbonate (DEPC) is a chemical which prevents the growth of microbes in fruit drinks and alcoholic beverages. It has been more widely used in Europe than in the United States. Unfortunately, it has been found to combine with ammonia to form a potent carcinogen called urethan. Although it is still used by a few American manufacturers, it is never listed on the label. This is because the DEPC itself breaks down before the drink reaches the consumer—which, however, does not rule out the presence of small amounts of urethan. (In addition, manufacturers are not required to list the ingredients in alcoholic beverages.)

EDTA—EDTA is used to prevent fat and oil products from going rancid. It is contained in many foods, especially salad dressings, oleomargarine, sandwich spreads, and mayonnaise. It is also used to prevent processed fruits and vegetables from turning brown. EDTA works by trapping metal impurities which may be present in the food (in fact, physicians treat metal poisoning with EDTA injections).

When present in excess, EDTA can cause kidney damage, calcium imbalance, and a condition resembling a vitamin deficiency. When used improperly intravenously, it can cause the teeth to dissolve!

Monosodium glutamate (MSG)—This is well known to most of us. In 1908, a Japanese chemist named Dr. Kikunae Ikeda discovered that MSG was able to greatly intensify the flavors of protein foods. Since then, it has become widely used around the world. The first indication that it posed any dangers came when a doctor named Ho Man Kwok discovered what he called the "Chinese Restaurant Syndrome." Dr. Kwok found that after eating in a Chinese restaurant he experienced headaches, tightness in his chest, and burning sensations in his forearms and at the back of his neck. The cause was finally traced to the MSG, especially the large amounts used in the soups.

In 1969, Dr. John Olney of the Washington University School of Medicine, St. Louis, found that large amounts of MSG fed to infant mice destroyed nerve cells of the hypothalamus—the area of the brain which controls the appetite and other important bodily functions. MSG has also been found to damage the retina of infant rats and mice (the human eye, being more developed at birth, is less susceptible to such damage).

Until October, 1969, most baby food manufacturers were putting MSG in their products, in order to make them more appealing to the parents. Public pressure finally forced them to cease this practice. There is some question among scientists whether large amounts of MSG could have adverse effects on fetuses. It is recommended that pregnant women restrict their intake of MSG.

Propyl gallate—Propyl gallate is an antioxidant which is often used in combination with BHT and BHA. In a study reported in 1965, researchers found that high levels of propyl gallate caused damage to the mitotic (cell division) mechanism of the liver cells. Further and more extensive research is indicated, especially as this is a widely used additive.

Saccharin—Most of us are familiar with this sweetener. It is used in many low-calorie products, and used by many persons in tablet form as a sweetener for coffee and tea. It is an important aid to diabetics, as the body does not convert it to glucose, as it does sugar. Although saccharin has not been shown to endanger human health, there is some cause for caution. In 1951, an FDA study found that high levels of saccharin caused kidney damage in rats, but did not produce cancer. Since then, however, several studies have pointed to the possibility that saccharin may cause cancer of the bladder. Therefore, it may be wiser for non-diabetic persons to limit their use of saccharin-sweetened foods—if they feel they need to use saccharin at all.

Polyvinyl chloride—Vinyl chloride, a packaging material, is not technically a food additive. However, it becomes one when small amounts of it migrate to the food inside containers. The Society of Plastics Industry, individual manufacturers, and foreign governmental agencies have produced evidence showing that vinyl chloride is able to migrate to different types of foods, including acidic, fatty, and alcoholic foods.

The Delaney Amendment

The Delaney Amendment to the Food and Drug Act, which was authored by Rep. James J. Delaney over twenty years ago, states that any additive which has been shown in tests to cause cancer must be banned. The precise wording is:

"No additive shall be deemed to be safe if it is found to induce cancer when ingested by man or animal, or if it is found, after tests which are appropriate for the evaluation of the safety of food additives, to induce cancer in man or animal."

The basic premise of this rule is that there is no tolerance, or "safe," level for such substances. It is this amendment which opened the way for the banning of cyclamates, for without it, the F.D.A. could have merely lowered the level of cyclamates allowed in foods.

Naturally, a piece of legislation as sound as the Delaney Amendment is bound to find enemies. Various self-interest groups have tried, and are still trying, to have this amendment repealed in order to allow small amounts of cancer-causing additives in foods. Many livestock producers, for example, would like the Food and Drug law to be changed to allow small amounts of diethylstilbestrol (DES) in meat. Yet DES is a proven carcinogen; a number of teenage daughters of women who took DES in pregnancy have died from an extremely rare form of cervical cancer.

Some scientists argue that as more sophisticated detection instruments are developed, more and more trace contaminants will be found in our food, and eventually all foods will have to be banned. This argument greatly distorts the basic idea of the Delaney Amendment, for the amendment applies only to intentional additives, over which manufacturers have control. DDT and other pesticides, for example, do not fall into this category of intentional additives.

Nonindustry supported scientists almost invariably agree with the principle that there is a zero safe level of cancer-causing substances. Those groups that advocate allowing small amounts of carcinogens in foods tend heavily to reflect the profit (not health) interests of industry. The National Academy of Sciences' Food Protection Committee, for example, is independently financed by grants from commercial laboratories; chemical, food, and packaging companies; and individuals.

In addition to the previously mentioned additives in this chapter, the following short listing of additives will further explain the most commonly used ones. Since there are thousands of them, the list must necessarily be incomplete.

For the reader who is interested in a more complete compilation of food additives, we suggest *A Consumer's Dictionary of Food Additives*, by Ruth Winter (Crown Publishers, New York,) and Beatrice Trum Hunter's *Food Additives And Your Health* (Keats Publishing, New Canaan, Conn.). These two works list thousands of additives and describes them in detail.

Agar-agar (or Japanese isinglass)—Used as a stabilizer and thickener. It is transparent, tasteless, and odorless. Agar-agar is obtained from several kinds of seaweeds that grow in the Pacific and Indian Oceans and in the Sea of Japan. It is used in ice cream, ices, frozen custard, sherbet, beverages, jellies and preserves, meringue, baked goods, icings, candied sweet potatoes, and confections; may be used as a substitute for gelatin; and may be a thickener in milk and cream. Nontoxic. GRAS

Ascorbic acid (Vitamin C)—The antiscurvy vitamin; it is essential for healthy bones, teeth, and blood vessels. It also is a preventive and remedy for the common cold. Ascorbic acid is believed by some authorities to clear cholesterol deposits from the arteries. It is added to fruit juices, frozen and concentrated fruit drinks, orange and lemonade, and carbonated drinks. It is used as an antioxidant and preservative in frozen fruits (particularly peaches), frozen fish dips, dry milk solids, fluid milk, beer and ale, apple juice, candy, soft drinks, artificially sweetened jellies and preserves, canned mushrooms, and flavoring oils. It is put into the pickling liquid in which beef and pork are cured; also used in packing pulverized, cured, and cooked meat products. Ascorbic acid may be toxic when ingested along with sodium salicylate, sodium nitrate, theobromine sodium salicylate, and methenamine; also barbiturates.

Benzaldehyde (or benzoic aldehyde)—An essential oil

derived from almonds or made synthetically. Occurs naturally in almonds, cassia bark, bitter oil, cajeput oil, tea, raspberries, cherries. It is used in apricot, brandy, rum, peach, liquor, cherry, berry, butter, coconut, pistachio, almond, pecan, vanilla, and spice flavorings. Also in ice cream, candy, ices, beverages, cordials, baked goods, and chewing gum. It may cause a skin rash and it is narcotic in high concentrations. Large amounts produce convulsions and central nervous system depression. The fatal dose is approximately two ounces. GRAS

Butylated hydroxyanisole (BHA)—A white or slightly yellow waxy solid. It is a preservative and antioxidant widely used in breakfast cereals, baked goods, potato flakes, sweet potato flakes, dry mixes for desserts and beverages, shortenings, lard, dry yeast, soup bases, chewing gum, ices, glacéed fruits, beverages, and unsmoked dry sausage. Allowed up to 50 parts per million (ppm) with BHT (*which see*) in dry cereals; 50 ppm in potato flakes; 1000 ppm in dry yeast; up to 0.02 percent of fat and oil content of food. Some suspected harmful effects which the F.D.A. is investigating.

Butylated hydroxytoluene (BHT)—A white crystalline solid. It is an antioxidant used in white potato and sweet potato flakes and in dry breakfast cereals; a chewing gum base, and emulsion stabilizer for animal fats and shortenings containing animal fats. Also used as an antioxidant to retard rancidity in frozen fresh pork and freeze-dried meats in amounts up to 0.01 percent of the fat content; allowed as 50 ppm in dry breakfast cereals and potato flakes; up to 200 ppm in emulsion stabilizers. Experiments with rats have indicated harmful effects. BHT is under investigation by the F.D.A, but is on its GRAS list. BHT is banned in foods in England.

Carrageenan chondrcus extract (or simply

carrageenan)—A glue-like derivative of Irish moss having
a seaweedlike odor and a salty taste. It is used as an
emulsifier and stabilizer in chocolate products, chocolate
milk, chocolate-flavored drinks, pressure-dispensed
whipped cream, confections, syrups for frozen products,
evaporated milk, cheese spreads, ice cream, sherbets,
frozen custard, ices, French dressing, and artificially
sweetened jellies and jams. Ammonium, calcium, sodium,
and potassium salts of carrageenan are used in syrups,
jellies, puddings, baked goods, and beverages. Also used
in medicinal syrup for soothing mucous membrane
irritation. Sodium carrageenan is in the F.D.A.'s list for
study for mutagenic, teratogenic, reproductive, and
subacute effects, but is on the GRAS list.

Citric acid—A varied additive used in the United States
for almost a century. It is a colorless or white crystalline
solid, having a strong acid taste. It is found in plant and
animal tissues and fluids, and is produced commercially
from the fermentation of crude sugar, or by extraction
from citrus fruits. Besides occurring naturally in citrus
fruits, it is found in peaches and coffee beans. It is used to
adjust the acid-alkaline balance of candies, jams, jellies,
wines, fruit juices, canned fruit, carbonated beverages,
frozen fruit, canned vegetables, sherbet, cheese spreads,
frozen dairy products, confections, dried egg white,
mayonnaise, salad dressing, canned figs, fruit butter,
preserves, fresh beef blood. Citric acid is also used in
curing meats, firming peppers, potatoes, tomatoes, and
lima beans; and to prevent off-flavors in potato chips and
French-fried potatoes. It imparts a fruit flavor to
beverages (2500 ppm), candy (4300 ppm), chewing gum
(3600 ppm), ice cream, and ices. Medicinally, it has been
used to dissolve urinary bladder stones. Non-toxic. GRAS

Citrus Red No. 2—A dye used for coloring orange skins of

oranges *not* intended for processing, and which meet minimum ripeness standards established by the laws of states in which the oranges are grown. Oranges colored with Citrus Red No. 2 must not bear more than 2 ppm, calculated on the weight of the whole fruit. The toxicity of Citrus Red No. 2 is not entirely determined. One constituent of the dye, 2-naphthol, if ingested in quantity can cause clouding of the eye lens, vomiting, kidney damage, and circulatory collapse. Application of this constituent to the skin can cause peeling and death if it is on a large area.

Dextrin (or British gum, starch gum)—A white or yellow powder made from starch. Used as a foam stabilizer for beer (causing the foam to last longer), a diluting agent for pills and dry extracts; for thickening industrial dye pastes; in matches, explosives, and fireworks; in dry bandages; for preparing emulsions; and in polishing cereals. Non-toxic. GRAS

Diethylstilbestrol (DES, or stilbestrol)—A synthetic estrogen fed to cattle and poultry (prior to the autumn of 1972) to fatten them. DES is a known carcinogen. In 1971, it was shown that daughters of women who had taken DES during pregnancy developed a rare form of cancer. Cattle fed DES were supposed to be taken off DES feed for a length of time that allowed the estrogen to be eliminated from the animals' bodies. However, tests showed residues of DES in refrigerated meat. Since there may be a lapse of two years between slaughter and sale of meat, diethylstilbestrol is still to be found in meat from cattle fed this hormone. The European Common Market, Sweden, and Italy have banned the use of DES in cattle.

Disodium guanylate—A flavor intensifier. It is the disodium salt of 5'-guanylic acid, found throughout nature as a compound in the formation of RNA and DNA. It is obtained from certain kinds of mushrooms. It is not known to be toxic.

Disodium phosphate—A sequestering agent. It is used in evaporated milk (up to 0.1 percent of finished product); in macaroni and noodle products (0.5 to 1.0 percent). It is used as an emulsifier in cheeses (3 percent by weight). It has U.S. Department of Agriculture clearance for use in preventing cook-out juices in cured pork shoulders and loins; cured, canned, and chopped hams; and bacon. Disodium phosphate is also used as a buffer in adjusting acidity of chocolate products, beverages, enriched farina, and sauces and toppings. Medicinally, it has uses as a cathartic, purgative, and in phosphorus deficiency treatment. It can irritate the skin and mucous membranes. GRAS

FD and C colors (food, drug and cosmetic colors)—These are color additives, which means that they are dyes, pigments, or other substances capable of coloring a food or a drug or cosmetic used anywhere in or on the human body. Legislation in 1938 required that all the additive colors be given numbers and that every batch be certified. Illnesses caused by some of the colors led to their being taken off the approved list. In 1959, the F.D.A. approved the use of "lakes," which are the dyes mixed with calcium or aluminum hydrates. The result is an insoluble coloring material used in candies and to color egg shells. In 1960, a law was passed to require tests for determining the suitability of all colors prior to putting them on a permanent listing as acceptable. Orange B (150 ppm), for coloring sausage casings, and Citrus Red No. 2 (2 ppm) for dying orange skins are now permanently listed. Red No. 3, Yellow No. 5, Blue No. 3, and Red No. 40 are also on the permanent list without any restrictions. All other colors are on the temporary list. The World Health Organization lists as competely acceptable only Yellow No. 5, Yellow No. 6, Blue No. 1, and Green No. 3.

Glycerides (monoglycerides, diglycerides, and monosodium glycerides of edible fats and oils)—These are emulsifying and defoaming agents. They are used in bread and other baked goods to maintain softness (which they do by absorbing moisture from the air), in ices, ice cream, ice milk, milk, beverages, shortenings, lard, oleomargarine, chewing gum base, confections, chocolate, sweet chocolate, whipped toppings, and rendered animal fats. The diglycerides are on the FDA list for study as possible mutagenic, teratogenic, subacute, and reproductive effects, but are on the GRAS list.

Inosinate—A flavor intensifier made from inosinic acid. It is obtained from meat extract or dried sardines. Non-toxic.

Lecithin—This substance is found in all living organisms, both plant and animal, as a constituent of nerve tissue and brain material. It is made up of choline, phosphoric acid, glycerine, and fatty acids in chemical combination. It is obtained commercially from eggs, corn, and soybeans. It is used as an antioxidant in breakfast cereals, candy, bread, rolls, buns, and other baked goods, sweet chocolate, and oleomargarine. *Hydroxylated lecithin* is a defoaming agent used in the production of yeast and beet sugar. Non-toxic.

Menthol—A flavoring agent that occurs naturally in mints, raspberries, and betel nuts. Its chief source is oil of peppermint. It is used in caramel, fruit, butter, spearmint and peppermint flavorings in chewing gum, candy, ices, ice cream, baked goods, liqueurs, and beverages. It is used in perfumery, and medicinally in cough drops and nasal inhalers; also in cigarettes. In large doses it can cause severe abdominal pain, nausea, vomiting, vertigo, and coma. The lethal dose in rats is 2.0 grams per gram of body weight. GRAS

Nitrate (potassium and sodium)—Potassium nitrate is also known as saltpeter and niter. In concentrations up to 0.02 percent, it is used as a color fixative in cured meats, in pickling brine (7 pounds per hundred gallons), and in chopped meat, such as hamburger (2.75 ounces per 100 pounds). Sodium nitrate, also called Chile saltpeter, is used as a color fixative in cured meats, frankfurters, bologna, bacon, uncooked smoked ham, poultry, wild game, meat spreads, Vienna sausages, potted meat, poultry, and spiced ham. Also used in baby foods. Nitrates combine with substances—secondary amines—found naturally in the stomach and in certain foods, producing powerful carcinogenic substances—nitrosamines. The combination of nitrates and certain stomach chemicals produces nitrites that have caused deaths of babies through the condition called methemoglobinemia, which cuts off oxygen from the brain. The F.D.A. has given nitrates priority in testing for cancer-causing effects, and also mutagenic, teratogenic, subacute, and reproductive effects.

Nitrite (potassium and sodium)—Potassium nitrite (up to 0.02 percent) is used to fix color of cured meats. It reacts chemically with the myoglobin molecule of meats, giving them a red-blooded color, brings a tangy taste, and resists the growth of the spore of *Clostridium botulinum*, which is responsible for the almost always fatal botulism of spoiling meat. Nitrite is a color fixative in bacon, bologna, deviled ham, cured meat, potted meat, meat spreads, Vienna sausage, spiced ham, smoke-cured tuna, shad, and salmon. It is used as a cosmetic and in baby foods. As with nitrate, nitrite can combine with certain stomach and food chemicals—secondary amines—to form the cancer-causing agents, nitrosamines and nitrosamides. Deaths due to nitrites are known, and the F.D.A. is giving nitrites the same priority testing that it is giving to nitrates. Meanwhile the Department of Agriculture is holding use of nitrite to 200 ppm going into ready-to-eat meats, which lose some nitrite en route to the table.

Pectin (including low methoxyl pectin and sodium pectinate)—Pectin is found in plant roots, stems, seeds, and fruits, where it acts as a cementing agent. It is a powder (coarse or fine) with a gluey taste, and almost odorless. The richest source of pectin is orange or lemon rind, which contains about 30 percent. It is used as a thickener, bodying agent, or stabilizer for artificially sweetened beverages, ice cream, ice milk, water ices, fruit sherbets, syrups for frozen products; fruit jelly, preserves, jams to make up for a lack of natural pectin; French dressing. Medicinally, it is an anti-diarrheal agent. Nontoxic.

Potassium citrate—A white or transparent powder, odorless, and having a cool salty taste. Used as a buffer in artificially sweetened jellies and preserves and in confections. Medicinally it is a gastric antacid and urinary alkalizer. GRAS

Sodium bisulfite (or sodium acid sulfite, or sodium hydrogen sulfite)—A white powder used as a bleaching agent in ale, beer, wine, and other food products. The F.D.A. is studying it for possible mutagenic effects, but it is on the GRAS list.

Sodium citrate—White odorless crystals, powder, or granules, having a salty taste. Prevents "cream plug," the semisolid collection of fatty material at the top of a container of cream. Also prevents "feathering" when cream is put into coffee. It is used as an emulsifier in evaporated milk, ice cream, processed cheese; as a buffer for controlling acidity and preventing loss of carbonation in soft drinks; also used in fruit drinks, confections, jams, jellies, and preserves. GRAS

Sorbitol—A white, crystalline powder with a sweet taste. It is found in berries, cherries, plums, pears, apples, seaweed, where it occurs due to the breakdown of dextrose. It is used as a sugar substitute for diabetics, but its safety for this use has not been proved. It is also used as a sequestrant in vegetable oils, a thickener in candy, a stabilizer and sweetener in frozen desserts used in special diets, as a humectant (retainer of moisture) and texturizing agent in soft drinks, shredded coconut, and dietetic fruits. The F.D.A. has asked for further study of sorbitol.

The foregoing list was chosen from thousands of food additives because it is made up of those most frequently found on food packages. We hope that you will refer to it when reading the make-up of the food within the packages you buy. Knowing something about the additives in the packaged foods may make you feel better and safer. Or, it may make you feel worse—in which case you will not buy that particular food anymore and thereby cut down on your body pollution.

References

CONGRESSIONAL HEARINGS

"Chemicals and the Future of Man," hearings before the Subcommitte on Executive Reorganization and Government

Research of the Committee on Government Operations, U.S. Senate, 92nd Congress, April 6 and 7, 1971 (chaired by Sen. Abraham A. Ribicoff).

"Cyclamate Artifical Sweeteners," hearings before a subcommittee of the Committee on Government Operations, House of Representatives, 91st Congress, 2nd session, June 10, 1970 (chaired by Rep. L. H. Fountain).

"Nutrition and Human Needs—Food Additives," Parts 4A, 4B and 4C, hearings before the Select Committee on Nutrition and Human Needs, U.S. Senate, 92nd Congress, 19, 20, 21, 1972 (chaired by Sen. Gaylord Nelson).

"Regulation of Diethylstilbestrol (DES)," hearings on S. 2818 before the Subcommittee on Health of the Committee on Labor and Public Welfare, U.S. Senate, 92nd Congress, 2nd session, July 20, 1972 (chaired by Sen. Edward M. Kennedy).

"Regulation of Diethylstilbestrol (DES)," Parts I and II, hearings before a subcommittee of the Committee on Government Operations, House of Representatives, 92nd Congress, Nov. 11 and Dec. 13, 1971 (Fountain).

"Regulation of Food Additives and Medicated Animal Feeds," hearings before a subcommittee of the Committee on Government Operations, House of Representatives, 92nd Congress, March 16, 17, 18, 29 and 30, 1971 (Fountain).

"The Toxic Substances Control Act of 1971 and Amendment," hearings on S. 1478 before the Subcommittee on the Environment of the Committee on Commerce, U. S. Senate, 92nd Congress, 1st session, Aug. 3, 4, 5, Oct. 4 and Nov. 5, 1971.

REPORTS

"Cancer Prevention and the Delaney Clause," a report by the Health Research Group, 2000 P. Street, N.W., Washington, D.C., January 1973, funded by Ralph Nader's Public Citizen.

"The FDA and Nitrite," a case study of violations of the Food, Drug and Cosmetic Act with respect to a particular food additive, mimeographed report by Dale Hattis, Dept. of Genetics, Stanford University School of Medicine, April 25, 1972, sponsored and distributed by the Environmental Defense Fund, Washington, D.C.

"How Sodium Nitrite Can Affect Your Health," report by Michael F. Jacobson, Ph.D., Center for Science in the Public

Interest, 1779 Church Street, N.W., Washington, D.C. 20036, February 1973 ($2).

"Regulation of Food Additives—Nitrites and Nitrates," 19th report of the Committee on Government Operations, House of Representatives, Aug. 15, 1972, based on a study by the Intergovernmental Relations Subcommittee (a Fountain committee report).

NEWSLETTERS

Food Chemical News, ed. by Lou Rothschild, 1341 G Street, N.W., Washington, D.C. 20005.

Newsletter of the Federation of Homemakers, Inc., ed. by Mrs. Ruth Desmond, 927 North Stuart Street, Arlington, Va. 22203.

Caveat Emptor, ed. by Robert Berko. Box 336, South Orange, New Jersey 07079.

Who Is Responsible?

We have described body polluting practices in the food, drug, cosmetics, fertilizer, and pesticide industries, and also among cattle, hog and poultry raisers, and crop growers. These practices should leave little doubt that all Americans daily face a serious problem of eating a healthful diet while having to buy foods polluted with scores of chemical additives. Of not knowing which prescription or nonprescription drugs may do far more harm than good or may be simply worthless. Of risking healthiness of skin and hair when using cosmetics that can actually damage the parts of the body they are meant to enhance.

The worst aspect of this problem is that the consumer is almost helpless against those whose eagerness for profit presents him with the choice of body-polluting items in the supermarket or drug counter.

The Big Food Industry

Nearly 150 billion dollars' worth of food, detergents, and cosmetics are sold by food stores each year, according to an estimate by the Food and Drug Administration. This is about one-fifth of all the goods and services sold in the United States each year. The very rich food industry can and does afford a powerful public relations apparatus and lobby.

In 1960, Pulitzer Prize-winning ex-reporter William Longgood wrote *The Poisons in Your Food*, one of the first books to expose the dangers in food processing and in agriculture. Shortly after the book was published, checkers at most supermarkets were given pamphlets to put into one of the bags of every customer, along with the groceries and sales slips. Entitled "The Good in Your Food," the pamphlet assured supermarket shoppers that all was well as far as their health was concerned. The food in the brown paper sack was healthful and harmless. Additives were put into food to make it taste better, keep better, and look better, just as every shopper really wanted it to be.

The pamphlet came from the public relations staff of the Supermarket Institute. This group of executives, meeting in their 23rd annual convention, had felt that Longgood's book and other similar rumblings of discontent made public reassurance necessary. One supermarket executive said, "Our entire economy and way of life is based on faith. If faith in the wholesomeness of our food is undermined, it can have a serious effect on the health of our nation. It can seriously affect the public's confidence in government, agriculture, science, and education, as well as in manufacturers, processors, and distributors of food." To stem the possibility of this kind of national discontent, the pamphlet was written, printed, and stuffed into brown paper bags.

The Food Association

There are a number of organizations encompassing most of the members of the food industry. These are nationwide groups with much money and a great amount of lobbying power. Among them are the American Meat Institute, National Livestock and Meat Board, American Dairy Association, National Dairy Council, Wheat Foods Association, Wheat Flour Institute, Cereal Institute, and

Sugar Research Foundation, Inc. The words "institute," "board," "council," and "foundation" give these public relations and lobbying organizations an odor of respectability.

The Sugar-Coated Bamboozle

Almost all of these organizations carry on some research into the uses of the particular kinds of food each represents. They also subsidize research in universities and research institutes. It is hard to say how much this research benefits any part of the population besides its sponsors, but one thing certain is that research which turns up findings unfavorable to sponsors conceivably could be downplayed.

The sugar industry provides an example. These trade organizations were founded in the early 1900s by many sugar producers and refiners of cane and beet sugar in the United States and other countries. Since its foundings, its membership has more than doubled. One of its publications explained one of its research purposes thus: "The purpose of our dental caries research is to find out how tooth decay may be controlled effectively without restriction of sugar intake." In the industry's search for an answer, it supported work by James H. Shaw and associates, biochemists, for more than ten years. At the Harvard School of Dental Medicine, the researchers explored the effects of sugar on the teeth of laboratory animals. One population of rats was fed a sugar-rich diet, while a control group was put on a sugar-free regimen. The results were what one might expect—the control group showed almost no cavities, but the rats on the sugar diet suffered many. From these results, Dr. Shaw reported that "we should cut down on our sugar consumption, particularly candy. We should be careful about sugars that remain in the mouth because of their physical properties."

This report ended the research project because it prompted the sugar sponsors' financial withdrawal. The results of the experiments were published in the *Journal of the American Medical Association,* but never appeared where the public would be likely to see them.

The Sugar Research Institute spends large amounts of money on promoting the idea that sugar gives one quick energy. The industry also gives aid to soft drink bottlers who are trying to neutralize the efforts of an increasing number of dentists who warn their patients that beverages containing sugar are not good for the teeth. Furthermore, the industry urges canners to use 60 percent more sugar "to gain maximum consumer acceptance," a successful campaign that has resulted in the difficulty of getting—except in health food stores—fruits canned in their own juices, rather than the thick syrups used by the large canners.

The Nutrition Foundation, Inc.

In 1941, the leaders of the food industry formed the Nutrition Foundation, Inc. Companies that were director-members included American Can, American Sugar Refining, Beechnut, California Packing, Campbell Soup, Coca-Cola, Container Corporation, Continental Can, Corn Products Refining, General Foods, General Mills, H. J. Heinz, Libby, McNeill, & Libby, National Biscuit, National Dairy Products, Owen-Illinois Glass, Pillsbury Flour Mills, Quaker Oats, Safeway, Standard Brands, Swift, and United Fruit.

Sustaining members included Abbott's Dairies, American Home Foods, American Lecithin, Bowman Dairy, Continental Foods, Cross & Blackwell, Curtiss Candy, R.

B. Davis, William Davis, Drackett, Flako, Gerber, Golden State, Hansen Laboratories, Knox Gelatine, McCormick, Minnesota Valley Canning, National Sugar Refining, Nut & Chocolate, E. Prichard, Red Star Yeast, Stouffer, Weston, and Zinsmaster. Financial contributions came from American Maize Products, A&P, Hawaiian Pineapple, Eli Lilly, Merck, Penick and Ford, and A. E. Staley.

We have risked boring you with this long list of companies in order to give you an idea of what the contributing resources of the Nutrition Foundation, Inc. are, and what lobbying power they can wield. The Foundation sponsors research into many aspects of nutrition through grants to medical and dental schools and other university departments. It cannot be denied that some valuable findings have come out of this research. However, we would like to see more consumer-oriented individuals and companies associated with it.

Science in the Service of Whom?

One of the largest recipients of money from the food industry is the Department of Nutrition at Harvard University, headed by Dr. Frederick J. Stare. Dr. Stare has impressive academic qualifications and his administrative ability is obvious. Yet, allegedly, he is the author of the following statements:

"Sugar is a quick energy food.... Even people on a severe reduction diet can afford to put a teaspoonful of sugar in their tea or coffee three or four times a day."

"The nutritive qualities of canned evaporated milk are every bit as good as those of fresh pasteurized milk." (In April, 1964, Dr. Stare was elected a member of the board of directors of the Continental Can Company.)

An after-school snack for teenagers attributed to Dr. Stare is "iced tea, lemon- or limeade, or Coke."

These statements should lead one to wonder whether Dr. Stare has the time to do much reading in his own field, since there is a considerable amount of sound research that contradicts his pronouncements.

Dr. Stare is not the only scientist who serves a dual role as a consultant of the food industry and as a researcher in the academic world.

One of the results of industry-sponsored research is the existence of a number of apologists for food industry processing practices. Dual-role scientists are continually testifying at Congressional hearings and are almost invariably found pooh-poohing the possible dangers of food additives and other adulterants, as well as nutrition-decreasing processing practices.

That the same lobbying groups are not in existence only to encourage research in nutrition is easily seen in their public relations activities, which expend more than a million dollars a year. They send "informational" stories on nutrition to newspapers. Also, editorial material is sent to trade journals and consumer magazines which have an estimated total readership of more than 50 million.

The project, Dial-a-Dietician, which was begun in Detroit in the late fifties, was launched with money from the food industry. Any consumer who dials an advertised phone number is given nutritional information—information that assures him the food available at the usual markets is wholesome, nutritional, and safe.

When William Longgood's *The Poisons in Your Food* was published, Dr. William J. Darby, a nutritionist, led criticism on the book. We do not feel Dr. Darby's criticisms were accurate. We feel too much subjective emotion was present to present an objective and fair view of Mr. Longgood's book. Dr. Darby had also led the criticism on Rachel Carson's *Silent Spring*.

A continual flow of material, with the intent of helpful nutritional information, goes to newspapers, magazines, and trade journals. It is impossible for the nutritionally untrained reader to sift anything of worth from this flood of information, and even the trained reader must do much searching to straighten out the material that comes from the food industry.

The U.S. Government and the Food Industry

In 1958, Congress passed the Food Additives Amendment, Public Law 85-929. This law was the result of lengthy hearings conducted by Representative James J. Delaney. In 1949, when Rep. Delaney first announced these hearings, the Food and Drug Law Institute (FDLI) was formed. It had as sponsors just about all the members of the food industry who were involved in the nutrition industry.

The FDLI has on its roster of officials "public members" who came from government, especially those agencies charged with administering and enforcing the food laws. The officers of FDLI come from the food industries, and among them are "public trustees" like Dr. Darby, Dr. King, and other officers of the Nutrition Foundation.

The FDLI sponsors and pays for courses on food, cosmetics, and drug laws given in law schools because it wishes "to give sound advice to industry and government," so that "existing laws [can] be understood and observed and that amendments [can] be carefully considered and adopted only when sound and in the public interest."

Those invited to attend these instructions include

industry and government lawyers. This means, in the words of consumer advocate Beatrice Trum Hunter, that "the FDLI plays an important role in the drafting and interpreting of all legislation dealing with consumer protection in the areas of food, drugs, and cosmetics."

The FDLI also sponsors lectures, seminars, and conferences on food and drug laws. It publishes a journal and legal research texts, and maintains liaison with government agencies, universities, and domestic and international bar associations. Its information sources, legal skills, and technical knowledge are used at public hearings on food and drug laws. In short, some consumerists feel that FDLI is a wealthy, technically expert, widely connected lobby for the food and drug industry.

Whose Hats Are They Wearing?

As we learned in the section on MER/29 and Kevadon-Thalidomide, there is a very relaxed, fraternal relationship between some of the government regulatory agencies and the industries that the agencies are supposed to regulate. Officials are continually leaving government service to become highly paid executives in the industries they were formerly charged with regulating. Their associates who are zealous in guarding the public interest by strictly interpreting government regulations are not likely to end up with lucrative offers from industry.

Also, people from industry join government regulatory agencies for a time and then return to industry. The enthusiasm of such officials for guarding public interest at the possible expense of their former and future employers is uncertain.

There are many joint government-industry-public commissions and committees that act in an advisory capacity to government regulatory agencies or congressional committees. In these joint groups it is presumed that the industry members and government members will not act in any way that will hurt the interests of their respective agencies. This leaves only the public members for the position of the public's advocates. However, when the public members are closely scrutinized, a surprising number turn up as members of the foundations, institutes, and boards which are supported by industries.

Scientists for Hire

In the hearings before the House Select Committee to Investigate the Use of Chemicals in Food and Cosmetics (82nd Congress, 2nd session), Leonard Wickenden, an industrial chemist, testified:

"I have upon my desk an advertisement published by the National Fertilizer Association. I think it is fair to call it a typical advertisement. From beginning to end it is extremely biased. Its byline reveals that it was written by a distinguished professor on the staff of one of our more important agricultural colleges. If some of the professors in our agricultural colleges are employees of the National Fertilizer Association, or of any of its members, can we be quite confident that their teaching is entirely unbiased? If some of the members of the staffs of our state experiment stations are receiving compensation in any form from the same source, can we be fully satisfied that all their research work is wholly in the interest of the farmers and to no slightest degree in the interests of the manufacturers of poison sprays, or other materials or equipment used in agriculture?"

The point that Wickenden raises is not only valid when judging the work of academics in their own fields, but also concerns government because the faculty members of colleges and universities are so frequently to be found on governmental regulatory agencies' advisory boards. Also, they are among the most frequently called expert witnesses at congressional hearings. In the sciences many who teach or do research in universities are in some way connected with the National Academy of Sciences' National Research Council. This quasi-governmental agency is frequently called upon to perform tasks for the national government, such as appointing committees to review problems of public interest. For example, in 1966 the Department of Agriculture asked the NAS-NRC to appoint a Committee on Persistent Pesticides to investigate this problem, especially as it applied to DDT. More than 80 witnesses testified before this committee, nineteen of whom were ecologically oriented. Another nineteen were from the industries that made the hearings necessary in the first place. Three witnesses were from the food industry, fourteen from the public health field, and twenty-eight from agriculture (who planned agriculture's commitment to DDT!).

After hearing these witnesses, the committee did not recommend any kind of ban on DDT, but recommended further study of the problem.

There is a Committee on Pest Control and Wildlife Relationships within the NAS-NRC. It includes 43 "supporting agencies," of which 19 are chemical corporations and four are trade organizations. In 1962, this committee issued two reports. Upon reading them, Dr. Frank Egler, an ecologist, wrote:

"The problem of industries' influence on scientists who are on their payrolls as consultants, through research grants and otherwise, is a prickly one. It has been brought up in connection with these reports. My surprise is not that such influence exists, but that other scientists are so naive and unsophisticated as to refuse to believe it. The reader should at least know of such connections in appraising the final conclusions. In short, these two [reports] cannot be judged as scientific contributions. They are written in the style of a trained public relations official of industry, out to placate some segments of the public that were causing trouble. With different title and cover pages, they would serve admirably for publication and distribution by a manufacturers' trade associations. Indeed, they are being much quoted in such places." ("Pesticides and the National Academy of Scientists," *Atlantic Naturalist*, Oct.-Dec., 1962.)

Unnecessary Protection

In 1960, an article by Dr. E. V. Askey, a past president of the American Medical Association (AMA), appeared in the *Journal of the American Medical Association*. The article was titled, "Americans Wasting Millions on Vitamins." Dr. Askey said. "Americans have to go out of their way nutritionally speaking, to avoid being well-nourished." He went on to say that most citizens get all the vitamins they need in their daily diets, and that supplementing their daily food intakes with vitamins is a waste of money. Dr. Askey was strangely silent or unaware about concurrent nationwide surveys that showed large percentages of Americans of all ages and all economic levels to be eating diets deficient in one or more nutrients, according to the Recommended Daily Allowances published by the NAS-NRC's Food and Nutrition Board. Of course, the food industry grabbed Dr. Askey's statement and gave it the widest publicity.

Dr. Askey's statement was only one event in a continuing campaign against the use of vitamin (and mineral) supplements. In 1962, the late George P. Larrick, then the F.D.A. Commissioner, proposed regulations that would prohibit the sale or purchase of vitamins except by a doctor's prescription or in very small amounts. The F.D.A. did not contend that the vitamins were harmful, which would have been the only justification for the proposed regulations. Instead, they repeated Dr. Askey's general contention that the average American diet provides all the vitamins an individual needs and therefore purchase of supplementary vitamins is a waste of money. What Commissioner Larrick wanted to do was to force Americans to spend (or not spend) their money in ways he saw fit. He did not cite any statute that gave his agency this power.

He also did not explain how anyone would save money by first paying for a doctor's prescription and then buying the vitamins at prescription drug prices, which are about 3 times as much as when bought over the counter.

The F.D.A. went to the news media and gave them its views on vitamins and nutritional supplements, hoping to use the news media to win its case before any hearing on the proposed regulations were held. This move did not work because the news media allowed for counter opinions. Also, consumer groups, such as the National Health Federation, were keeping the public aware of the F.D.A.'s moves. As a result, Commissioner Larrick drew back and temporarily shelved his proposal.

The regulations to "protect" the consumer against buying any vitamins he might need remained on the shelf until Dr. James Goddard succeeded Larrick. Commissioner Goddard revised the regulations, tightening them by giving the F.D.A. even more control over the sale and purchase of vitamins and nutritional supplements. Every label that was to go on vitamins and dietary supplements was to bear the following:

"Vitamins and minerals are supplied in abundant

amounts by commonly available foods. Except for persons with special medical needs, there is no scientific basis for recommending routine use of dietary supplements." Originally, there had been more to the label: "The Food and Nutrition Board of the National Research Council recommends that dietary needs be satisfied by foods." This statement was deleted because the Food and Nutrition Board insisted that *it had never made such a statement* and that use of its name was unauthorized. Dr. LeRoy Voris, Secretary of the National Research Council, said that the Food and Nutrition Board had never reviewed nor approved the assertion attributed to it by the F.D.A.; what is more, there never was, as far as he knew, any discussion of that matter between the F.D.A. and the Board.

Other members of the Food and Nutrition Board also complained. For example, Dr. W. M. Sebrell, Jr., wrote, "This statement is objectionable and misleading, and uses the authority of the Food and Nutrition Board...to support a statement which, taken out of context, creates a false impression.

"The generalization that vitamins and minerals are supplied in abundant amounts in the food we eat has no relevance as applied to a particular individual." In other words, how does the F.D.A. know whether *you* get enough vitamins and minerals in what *you* eat?

Dr. Sebrell took exception to the part of the F.D.A.-proposed label that said "special medical needs." He said that "there must be many thousands of people in this country on restricted diets for the purpose of losing weight who do not have any serious medical need, but who should take vitamins because of limitations of their foods intake." And "there are certainly many thousands of people on special diets of other kinds for various reasons, medical and non-medical, for whom there would be a scientific basis for dietary supplementation."

Concerning the F.D.A.'s claim that Americans were getting all the vitamins needed in their daily diet, Dr. Thomas H. Jukes, lecturer in nutrition at the University of California, said, "I do not think any professor of nutrition would give a passing grade to a student who made such a statement."

After two years of behind-the-scenes foot-dragging, the F.D.A. opened hearings on the proposed regulation. Most of the witnesses were persons hand-picked by the government.

The F.D.A.'s attorneys expressed an aversion to having all sides of the question aired. In the proceedings, at least one focus of the F.D.A.'s bias against the health food industry emerged. Sidney Weissenberg, assistant associate Commissioner for Compliance, ranted against "so-called health food stores," "food faddists," and "quacks." Mr. Weissenberg spent 52 days on the witness stand—longer than any other witness. When he was told to desist from using the phrase "so-called health food stores," he ignored the examiner's order.

It turned out that he was a one-man board of censorship of what could and could not be printed on vitamin containers. "It is Mr. Weissenberg who first makes that determination," a government attorney testified. "F.D.A. relies on him to do this...every day with no guidelines ...that is what we bring our cases on." The F.D.A. attorneys showed Weissenberg a number of vitamin labels which had been handpicked to make the government's point, but it turned out that Weissenberg could not tell which were in current use.

One of the more vicious of the F.D.A.'s maneuvers at the hearing was to try to make it impossible for the opposition to get transcripts of the proceedings. The F.D.A. entered into a contract with the Columbia Reporting Company, of Washington, D.C., to buy the transcript at nine cents a page, but allowing the company to charge consumer groups and industry representatives 75 cents a page, a price most could not even come near paying.

Dr. Frederick J. Stare turned up in opposition to the F.D.A. He wrote the Examiner, "Strange as it may seem, we have little factual information on the present nutritional status of representative groups of people in our country." And, using the testimony of Surgeon General Dr. William Stewart, when he appeared before a Senate investigating committee in 1967: "We do not know the extent of malnutrition anywhere in the United States—I cannot say what the extent is, because we just don't know," Dr. Stare continued, "With such a clear statement from the head of our Public Health Service, a statement with which most physicians knowledgeable in nutrition would agree, I don't see how anyone at this time can intelligently recommend new regulations on special diet foods, vitamin and mineral fortified foods and supplements." Dr. Stare went on to urge that the hearings be recessed until more knowledge on the nation's nutritional status could be obtained. The hearings continued.

In January 1970, the new F.D.A. Commissioner Charles C. Edwards ordered the hearings ended in time to have a report ready by June 1. The Examiner announced that all testimony of opponents of the regulation had to be in by March 16. This meant that the F.D.A. had 16 months in which to present its testimony, but the opponents (104 of them) would have to rush through their case in three months.

The foregoing account of the F.D.A.'s attempt to restrict the over-the-counter sale of vitamins and minerals as food supplements has been described in some detail (regretfully leaving out a number of underhanded and even questionable actions on the part of the F.D.A.) because it is typical of the F.D.A.'s attitudes and actions in its handling of health-related matters that stem from its bureaucratic ideology. Most important, it shows the F.D.A.'s dictatorial methods used against those who disagree with that ideology.

When the long and expensive hearing came to an end in early 1971, the F.D.A. had not won the labeling victory it had sought; the results were inconclusive. But the F.D.A. does not give up easily. On December 12, 1972, the F.D.A. proposed restrictions on the potency of vitamins. The proposal was to limit vitamin A to 10,000 international units and vitamin D to 400 units per capsule or other unit dosage. The F.D.A. said the restrictions would apply to "vitamin products marketed as foods for special dietary purposes and as over-the-counter drugs."

The Associated Press dispatch said, "Americans have been taking vitamin supplements for decades, but concern has been heightened by the new health-food craze."

The F.D.A. announcement said that persons interested in the proposed restrictions have 60 days to comment on them. Also, these restrictions will be followed "in the near future" by broad regulations covering most other vitamins, minerals, and other food supplements.

The AP dispatch ended, "The F.D.A. is said to be considering rules setting minimum and maximum amounts of nutrients in multivitamins and mineral products, prohibiting health claims, and requiring expiration dates on labels." The AP writer did not know that this really was not news.

Perhaps one more example should be given. The account comes from the preface written by Dr. George W.

Crane to the book, *The Dictocrats' Attack on Health Foods and Vitamins*, by Omar V. Garrison. At a famous F.D.A. trial, a Justice Department attorney, who was not assigned to the case, asked one of the defendants, "Doctor, do you know why *you* were indicted? Especially since the government knew you were merely a neutral professor who had offered free advice as you would to any other scientist who might seek out your counsel. So why do you think you were indicted?"

"I've often wondered," the good doctor replied, "*why* did the government include me in the indictment?"

"Well, that's standard procedure," replied the Washington attorney. "The government knew that sooner or later you would be called into court as an expert witness for the defense, since you are the world's foremost physiologist and researcher.

"So the government indicted you to discredit any testimony you might make. It's part of our strategy to reduce the credibility of opposing witnesses!

"The average American thinks 'indictment' is synonymous with 'being guilty.' So our Department of Justice spreads its indictments over all possible witnesses for the defense!"

This horror tale, aside from being an account of something that one would expect in a totalitarian government, and not in the United States, should cause one to ask just what tactics like that on the part of the F.D.A. have to do with this agency's purpose, which is to protect the American citizen from unsafe drugs, dangerous additives and pesticide-engendered poisons in his food.

The News and Advertising Media

Since the time of Thomas Jefferson, the citizens of this

country have been told countless times that a free and impartial press is one of the main bastions of our liberty. Fortunately for us, and despite its many shortcomings, our press is free and doing a good—though far from perfect—job. Those who know something about the news media (of which, the press is, today, but one segment), are aware that the good things that emanate from the media are the work of a minority of the reporters and editors who are gathering, interpreting, and disseminating news. The rest follow the herd and do as little work as possible, and they are the main targets of news services that give information on events, policies, and persons in and out of government. What is important to us here is that reporters who rely on news services necessarily reflect the viewpoint of news services. And news services get much of their information from printed handouts.

The F.D.A., in its vendetta against health food makers, distributors, stores, and literature, grinds out scores of handouts castigating the whole health food enterprise as a huge crackpot venture. Thus, in the autumn of 1972, when, as we saw above, the F.D.A. opened another round in the Battle of the Vitamins, the Associated Press dispatch spoke about the "health food craze." It was easier for the Associated Press reporter to accept the stereotyped description of the health food movement than to find out something about it himself. This one instance is typical. The majority of the news media docilely goes along with the opinions of the health food movement that are held by the government agencies and the industry lobbies.

Omar Garrison tells of a raid made by F.D.A. agents and state food inspectors upon a small Detroit department store. The raiders seized a supply of safflower oil capsules, claiming that they were being used to promote the sale of a book, *Calories Don't Count*, from which the F.D.A.

wanted to "protect" the consumer. The *Detroit Free Press* ran a front-page story of the raid and accompanied the article with a four column photograph captioned, "Store employees look on as government agents seize 'reducing capsules.'" The newspaper story failed to give a reasonably equal account of the small department store's side of the raid.

In 1965, the then F.D.A. Commissioner James Goddard disclosed that his agency was distributing through a trade-union news agency "news articles" [handouts] giving the F.D.A.'s viewpoint on drugs and cosmetics. The stories were written by a free-lance writer working under contract to the F.D.A., not by any member of the working press. As printed in newspapers, the articles gave no clue that they were prepared for a government agency under its direction, and were not news stories at all.

The *Wall Street Journal* commented on this sneaky practice. "To be sure, it is not likely anybody has been done any serious harm by F.D.A.'s bashfulness about being identified as the source of this 'news' it is giving away free—the question is whether it is proper for any governmental agency to feed this propaganda in disguise to a segment of the press gullible enough to accept it."

In 1967, Princeton University Press published *The Medical Messiahs: A Social History of Health Quackery in Twentieth Century America,* by James Harvey Young. In the preface, Mr. Young stated that his research and writing had been supported over a span of years by a Public Health Service Research Grant. The grant also provided Mr. Young with a research assistant.

This practice is a double deployment. Public funds were being used to pay a writer to produce a book under the auspices of the government. The book, which reflects the viewpoint of its governmental sponsor, is then published under the imprint of a commercial publishing house. Thus, the first representation has been perpetrated. The taxpayer who buys the book has paid for its production with his tax dollars, then he pays for it a second time when he buys it. Thus, a second representation.

Using the F.D.A. Properly

The Food and Drug Administration is not a very large government agency as government agencies go, yet it has thousands of employees throughout the country. Most are in Washington, D.C., but a considerable number are spread out in many cities to make it convenient for the F.D.A. to carry on its work of inspecting food, drug, and cosmetic producers.

The need for the F.D.A. is obvious. With thousands of food additives available to food processors and more thousands of drugs and cosmetics on the market, and with scores more of these items continually being introduced, an impartial agency is needed to protect the public from poisonous, useless, and worthless foods, drugs, and cosmetics. The Food and Drug Administration is the governmental agency created by Congress to impartially and fairly carry out this work of protection.

The job the F.D.A. has been given is vast. There are more than 2,500,000 interstate shipments of fruits and vegetables each year in the United States. The F.D.A. is able to inspect only 1 per cent of this food. The reason is lack of inspectors.

The F.D.A. has to inspect about 90,000 food processing establishments for cleanliness and for the ingredients that go into the processed food. The F.D.A. has the manpower to inspect only one-fourth of the processors. That there is a need for such inspection is emphasized by the 80,000,000 pounds of food seized each year for filth, decomposition, and lack of sanitation on the premises.

Each year, nearly 400,000 shipments of foodstuffs, cosmetics, and drugs enter the United States from other countries, some lacking almost any health and sanitary laws.

A recent F.D.A. commissioner revealed that his agency receives, on an average working day, four applications for approval of drugs; eight proposals for the testing of new drugs; and 13 requests for modification of drugs previously approved. The processing of a single application may take almost 200 work days, involve 29 members of the F.D.A., six outside consultants, and 19 other outside contacts; also conferences, laboratory tests, and field trips. The record of this typical application may run to 4,000 pages.

Of the 14,000 drug establishments that the F.D.A. must inspect, it can get around to only half each year. The Commissioner said, "This is far from adequate coverage of an industry whose output is so closely associated with the health and well-being of every citizen." One result this situation causes is a great number of drug-induced deaths and permanent injuries due to drugs which are suffered by patients each year. We have seen some of them in the story of MER/29.

Every American citizen owes the Food and Drug Administration a debt, because most of its overworked staff do an honest and valuable job of protection. The competence and dedication of Dr. Frances Kelsey, who saved American women from the nightmare of Thalidomide-deformed children, are not exceptional in the F.D.A.

On the other hand, we have seen in this book how some employees are blindly prejudiced against the health food industry and command a large staff of agents and inspectors in a Gestapo-like war of extermination on the objects of their prejudice. We have seen, in the case of the vitamin hearings, how the F.D.A.'s legal staff can engage in patently dishonest tactics that deny some of the basic freedoms that are the right of every American citizen. How readily and eagerly the F.D.A. will take destructive confiscatory action against a small health food store for allegedly mislabeling a few bottles of honey, but then do nothing about confiscating several hundred thousand bottles of wine mislabeled by a large vintner. It is these things—very, very serious things—that give a useful and desperately needed agency a bad reputation.

Since the Food and Drug Administration has so important a part to play in the life of every citizen, all citizens should be concerned with making the F.D.A. work properly. Unfortunately, most citizens are ignorant of the problem and many who have any knowledge of it are indifferent. The burden of taking action then falls upon a few thinking, concerned citizens. That the probability of reforming the F.D.A. is not as remote as it may seem can be seen in its changing position and character. Dr. Schmidt, its director, did not take the usual pro-industry, anti-consumer position which nearly all former directors of that agency have held. Instead, Dr. Schmidt has caused the agency to step up its investigation of suspected harmful food additives and drugs (causing nitrite and Red dye No. 2 to be banned in bacon). He has given a new spirit and sense of dignity to an agency that fell from grace with the American population. Let us hope that Dr. Schmidt remains strong and unflinching in his determination to keep our food and drug supply safe. Many consumerists are already comparing Dr. Schmidt's work with that of the creator of the first food safety agency, Dr. Wiley.

Taking Action

If what you have read so far upsets you, you are not alone. Millions of other American consumers are outraged by the pollution of our soil by pesticides and chemical fertilizers. They are frightened by the additives in the foods they and their children are forced to eat. Almost weekly, there are news reports of poisonings by food additives, or laboratory reports that another additive has been found to cause cancer. Consumers are angered by the bureaucratic high-handedness of the administrative and legal department of the Food and Drug Administration and other government agencies. As short a time as five years ago, the unhappy consumer could do little more than look on in frustration at the body-polluting situations that seemed to surround him. Today, he no longer needs to feel frustrated; there are constructive channels into which he can turn his anger and do something to fight body pollution.

Consumerism

The anti-pollution movement is a broad one. It includes action against not only those who pollute our bodies through foods but also through air filled with gaseous industrial wastes and automobile emissions, as well as lakes and streams fouled by solid and liquid industrial wastes and sewage. This is part of a broad ecological movement.

The anti-pollution movement is allied with those who

are tired of being cheated by large corporations that sell shoddy and unsafe merchandise, from children's toys to automobiles. This part of the movement is consumerism, and it includes action against the manufacturers of harmful and useless foods, drugs, and cosmetics.

The consumer movement has been a long time growing. It would be difficult to say just when this movement began. If a single event were to be picked as a beginning, it might be the publication in 1928 of a book titled *100,000,000 Guinea Pigs*, by Kallet and Schlink. This book made the reader aware of the great differences in the quality of the goods he was buying and how the manufacturers treated the average consumer as if he were a guinea pig on which to try out the sale of ill-made, harmful, or useless goods. From this book sprang a consumer goods testing organization, Consumers' Research, Inc., which published a monthly bulletin telling subscribers the results of the product tests. This enabled the consumer to exert a little influence by purchasing only the products that were reported to be healthful or well-made. In 1936, Consumers' Research split and a new testing organization, Consumers Union of the U.S., Inc., was founded. Today, the younger organization has become the larger, but both testing organizations are functioning and satisfying the consumer's desire for impartial product advice.

At first, the testing organizations were met with skepticism by many consumers, who believed that the testers were paid by manufacturers to certify products as being good buys, and the business community showed hostility to the product testers. But, these attitudes have largely been dispelled by the procedures used by the two research organizations: unannounced purchases of samples in the open market; refusal of permission to any company to use test results in its advertising; scientifically impartial testing procedures; allowing the members (subscribers) to elect the board of directors and to propose the products and services to be tested. Also, Consumers Union gives voice to its members' complaints against deceptive advertising, auto safety, and the functioning of governmental consumer protective agencies.

For a number of years, the actions and stands of the consumer testing organizations did not make much headway with legislators. One reason was that the country had not yet solved its "level of production" problems, and no legislator wanted to do anything to hinder sales of anything, and thereby slow down production. As the country moved to the solution of its production problems, legislators began to pay more attention to the voice of the consumer. In the fifties, Senators Estes Kefauver, Paul Douglas, Philip Hart, Warren Magnuson, and Abraham Ribicoff and Representative Wright Patman put consumer-oriented legislation into the hopper in unprecedented volume.

In the 1960's, the consumer movement came into its own. President Kennedy formed a Consumer Advisory Council and attached it to the Council of Economic Advisors. The council gave consumers a direct link to the White House. This led, under Johnson, to the appointment of a special presidential assistant for consumer affairs in the President's Committee on Consumer Interests. One result was the establishment of liaison between about 32 voluntary consumer organizations and the President's Committee. In a message to Congress in 1964, President Johnson said, "We cannot rest content until [the consumer] is in the front row, not displacing the interest of the producer, yet gaining rank and representation with that interest."

In the late 1960's, the Consumer Federation of America was formed as a coordinating organization for the efforts of consumer, labor, and agricultural, rural electrification, and purchasing cooperatives in promoting a consumer program in Congress. In 1969, President Nixon climbed aboard the consumerism bandwagon. He kept the position of special assistant for consumer affairs and sent Congress a "consumer bill of rights."

Meanwhile much consumer legislation had been passed. There is the Federal Insecticide, Fungicide, and Pesticide Amendment of 1964; National Traffic and Motor Vehicle Safety Act of 1966; Fair Packaging and Labeling Act of 1966; Wholesome Meat Act of 1967; Consumer Credit Protection Act of 1968; Truth in Lending Act of 1968; Child Protection and Toy Act of 1969; and the health hazard warning on cigarette packages (1968).

With all this legislation, why can food, drug, and cosmetics producers sometimes turn out the harmful, unwholesome, or useless products that still are being put on the market? An answer is that the laws themselves are not enough; strict enforcement is necessary. And it is in the realm of enforcement that the federal agencies fail the citizens they are supposed to serve. Also, no sooner is a law passed than lobbyists are on Capitol Hill pressuring congressmen to pass amending legislation that riddles the law with loopholes.

Political Action

It is a characteristic of most people that they can become politically aware and active if they feel personally harmed by some action or if a leader arouses their political awareness. They remain active until they achieve the goal that aroused them; then they become inactive. The lobbyists and politicians never rest because politics is their

business. To counter their efforts, you must be constantly
alert and active concerning the causes you are interested
in. This is easier said than done. Yet each citizen can be
very effective politically if he or she tries. You can join a
consumers group, such as Consumers Union. Your dues
will support their activities and they will listen to what you
have to say on consumer matters connected with the
problem of body pollution. They have their own lobbyists
in Washington; they will be lobbying for you. Then, you
can write to your representative and senators. Most people
feel that this is futile, but those who know Washington
know that letters from constituents are taken very
seriously by congressmen. Of course, they receive too
many letters to read each one; but their assistants do read
the letters and answer them. True, the answer may be a
form letter, but the contents of your letter were noted and
tabulated before the reply was sent to you. Enough letters
on a particular subject will make even the weakest
congressman stand up to the lobbyists. Make it a habit to
write a couple of letters every week. There is enough in
this book alone on the subject of body pollution to keep
you writing for years. How about starting right now by
writing your congressmen that the Food and Drug
Administration needs a shakeup, and why.

Economic Action

Economic action can be very effective. In a way, it is
what the whole matter of the consumer versus the profit-
hungry food processor or drug and cosmetic manufacturer is all about. If one of these producers finds that his
polluted product is not selling, he will stop putting it on the
market. After all, the only reason that adulterated and
harmful food, drugs, and cosmetics are sold is to make
money.

Again, we urge you to join some kind of consumer product-testing group, so that you can learn what you are buying. With knowledge of the products on the market, you certainly won't buy those that will harm you or are a waste of money.

You can be effective on the local level, too. Tell the merchants you deal with that you won't buy from them if they continue to sell products you know to be of a body-polluting nature. If you deal with a supermarket, get some friends together and picket the store. The manager will quickly listen to your complaints, and the supermarket's buyers will eventually hear about them.

Social Action

One of the most effective kinds of action is social. It is the kind that is taken by individuals upon seeing their friends, coworkers, or neighbors doing something that needs doing. Much of the work done to clean up the environment has been done by groups of individuals who formed because they felt a common need to do something. Simply setting a good example can be very effective. It could work this way:

If you have a lawn or garden, stop using pesticides and buy some black-light insect traps. Not only put them out, tell your neighbors about them and show how effective they are. A very large number of Americans are ecologically aware today. The insect trap idea could catch on. This would mean less pesticide draining into streams, rivers, and lakes, and less contamination of your drinking water.

The largest kind of green area in the United States is the house lawn. If you have one, show your neighbors how well it will grow with mulch instead of chemical fertilizer. Many neighbors will follow your lead. People really do want to cut down on pollution.

Legal Action

Another type of action you can take is legal action against the body polluters that you come up against every day when you buy and eat food. There are several ways to do this. If your pocket can afford it, you may take your complaint about a kind of food or cosmetic to a lawyer and let him handle the matter from there as a lawsuit. Very few of us are likely to do this. Instead, you can support one of the many organizations that employ legal staffs for the furtherance of consumers' rights and for the defense of the environment. One such organization is the Sierra Club, another is the Environmental Defense Fund, and there are others you can find if you ask environmentalists or consumer groups. The legal work is done by the organizations' legal staffs for you. All you have to do is help out financially through membership or a modest contribution. It was legal action by environmentalist groups that eventually led to the banning of DDT.

Most states and many local communities have consumer fraud divisions. By bringing your complaint to one of these, you can have legal action taken for you. You do not have to become ill from an adulterated food, for example, in order to take legal action against the processor. All you need to do is to show that the food violates some part of the Federal Food, Drug, and Cosmetic Act, the Wholesome Meat Act, or some other law or governmental regulation.

Section 402,b,1 of the Federal Food, Drug, and Cosmetic Act says: "A substance recognized as being a valuable constituent of food must not be omitted or abstracted in whole or in part . . ." Why not complain to a consumer fraud department, pointing out that enforcement of this part of the Act makes refined sugar and wheat minus wheat germ unlawful.

Section 402,b,2 says: ". . . nor may any substance be substituted for the food in whole or in part." Enforcement of this section would put an end to "enrichment" and "fortification" of foods that have had nutrients removed.

You can find in consumer protection laws dozens of other grounds for legal action to defend yourself against body pollution.

Also, one of the best and most respected consumer publications is *Caveat Emptor*, published monthly, Box 336, South Orange, New Jersey 07079. Many leading consumerists write for this excellent publication, including Ralph Nader and Dr. Michael Jacobson, who founded *Food Day* and heads the Center for Science in the Public Interest.

Plus:

Committee for Environmental Information
438 North Skinker Boulevard
St. Louis, Missouri 63130

Conservation Foundation
1250 Connecticut Avenue, N.W.
Washington, D.C. 20036

Consumers Union of U.S.
256 Washington Street
Mount Vernon, New York 10550

Defenders of Wildlife
1346 Connecticut Avenue, N.W.
Washington, D.C. 20036

Environmental Action
2000 P Street, N.W.
Washington, D.C. 20036

Environmental Defense Fund
1910 N. Street, N.W.
Washington, D.C. 20036

Friends of the Earth
30 East 42nd Street
New York, New York 10017

The Fund For Animals, Inc.
140 W. 57th Street
New York, N.Y. 10019

Izaak Walton League of America
1326 Waukegan Road
Glenview, Illinois 60025

National Audubon Society
1130 Fifth Avenue
New York, New York 10028

Scientists Institute for Public Information
30 East 68th Street
New York, New York 10021

Sierra Club
1050 Mills Tower
San Francisco, California 94104

Society for Social
   Responsibility in Science
221 Rock Hill Road
Bala-Cynwyd, Pennsylvania
   19004

Wilderness Society
729 15th Street, N.W.
Washington, D.C. 20005

Index